Memoirs of the American Mathematical Society
Number 310

Roger D. Nussbaum
and Heinz-Otto Peitgen

Special and spurious solutions of $\dot{x}(t) = -\alpha f(x(t-1))$

Published by the
AMERICAN MATHEMATICAL SOCIETY
Providence, Rhode Island, USA

September 1984 · Volume 51 · Number 310 (end of volume)

MEMOIRS of the American Mathematical Society

This journal is designed particularly for long research papers (and groups of cognate papers) in pure and applied mathematics. It includes, in general, longer papers than those in the TRANSACTIONS.

Mathematical papers intended for publication in the Memoirs should be addressed to one of the editors. Subjects, and the editors associated with them, follow:

Ordinary differential equations, partial differential equations and applied mathematics to JOEL A. SMOLLER, Department of Mathematics, University of Michigan, Ann Arbor, MI 48109.

Complex and harmonic analysis to LINDA PREISS ROTHSCHILD, Department of Mathematics, University of California at San Diego, LaJolla, CA 92093

Abstract analysis to WILLIAM B. JOHNSON, Department of Mathematics, Ohio State University, Columbus, OH 43210

Algebra, algebraic geometry and number theory to LANCE W. SMALL, Department of Mathematics, University of California at San Diego, LaJolla, CA 92093

Logic, set theory and general topology to KENNETH KUNEN, Department of Mathematics, University of Wisconsin, Madison, WI 53706

Topology to WALTER D. NEUMANN, Mathematical Sciences Research Institute, 2223 Fulton Street, Berkeley, CA 94720

Global analysis and differential geometry to TILLA KLOTZ MILNOR, Department of Mathematics, Hill Center, Rutgers University, New Brunswick, NJ 08903

Probability and statistics to DONALD L. BURKHOLDER, Department of Mathematics, University of Illinois, Urbana, IL 61801

Classical analysis to PETER W. JONES, Department of Mathematics, University of Chicago, Chicago, IL 60637

Combinatorics and number theory to RONALD GRAHAM, Mathematical Sciences Research Center, AT&T Bell Laboratories, 600 Mountain Avenue, Murray Hill, NJ 07974

All other communications to the editors should be addressed to the Managing Editor, R. O. WELLS, JR., Department of Mathematics, Rice University, Houston, TX 77251

MEMOIRS are printed by photo-offset from camera-ready copy fully prepared by the authors. Prospective authors are encouraged to request booklet giving detailed instructions regarding reproduction copy. Write to Editorial Office, American Mathematical Society, P. O. Box 6248, Providence, Rhode Island 02940. For general instructions, see last page of Memoir.

SUBSCRIPTION INFORMATION. The 1984 subscription begins with Number 289 and consists of six mailings, each containing one or more numbers. Subscription prices for 1984 are $148 list; $74 member. A late charge of 10% of the subscription price will be imposed upon orders received from nonmembers after January 1 of the subscription year. Subscribers outside the United States and India must pay a postage surcharge of $10; subscribers in India must pay a postage surcharge of $15. Each number may be ordered separately; *please specify number* when ordering an individual number. For prices and titles of recently released numbers, refer to the New Publications sections of the NOTICES of the American Mathematical Society.

BACK NUMBER INFORMATION. For back issues see the AMS Catalogue of Publications.

TRANSACTIONS of the American Mathematical Society

This journal consists of shorter tracts which are of the same general character as the papers published in the MEMOIRS. The editorial committee is identical with that for the MEMOIRS so that papers intended for publication in this series should be addressed to one of the editors listed above.

Subscriptions and orders for publications of the American Mathematical Society should be addressed to American Mathematical Society, P. O. Box 1571, Annex Station, Providence, R. I. 02901. *All orders must be accompanied by payment.* Other correspondence should be addressed to P. O. Box 6248, Providence, R. I. 02940.

MEMOIRS of the American Mathematical Society (ISSN 0065-9266) is published bimonthly (each volume consisting usually of more than one number) by the American Mathematical Society at 201 Charles Street, Providence, Rhode Island 02904. Second Class postage paid at Providence, Rhode Island 02940. Postmaster: Send address changes to Memoirs of the American Mathematical Society, American Mathematical Society, P. O. Box 6248, Providence, RI 02940.

The paper used in this journal is acid-free and falls within the guidelines established to ensure permanence and durability.

CONTENT

PAGE

Abstract

We consider periodic solutions of the parameter dependent differential-delay equation $\dot{x}(t) = -\alpha f(x(t-1))$ which exhibit special symmetries. Particular interest is in two major problems:

- existence and characterization of global α-continua of such solutions

- existence and explanation of spurious solutions.

The latter are solutions of numerical scheme which are by no means approximative solutions of the differential equation. Central methods and techniques for the first problem are new topological perturbation and phase plane like techniques. The second problem is analyzed on the background of associated homoclinic and fixed point structures.

AMS (MOS) Subject Classification 1980.
Primary: 34K15, 58F15
Secondary: 47H10, 47H15, 39A12, 58F22

Key words and phrases: differential-delay equations, periodic solutions, symmetries, numerical approximations, spurious solutions, homoclinic structures.

Library of Congress Cataloging in Publication Data

Nussbaum, Roger D., 1944-
Special and spurious solutions of $\dot{x}(t) = -\alpha f(x(t-1))$

(Memoirs of the American Mathematical Society, 0065-9266 ; no. 310 (Sept. 1984))
"Volume 51 number 310 (end of volume)."
Bibliography: p.
1. Functional differential equations--Delay equations--Numerical solutions. I. Peitgen, Heinz-Otto, 1945- . II. Title. III. Series: Memoirs of the American Mathematical Society ; no. 310.
QA3.A57 no. 310 [QA372] 510s [515.3'52] 84-14568
ISBN 0-8218-2311-6

0. INTRODUCTION

In this paper we study 'special periodic solutions' of

$$(0.1) \qquad \dot{x}(t) = -\alpha f(x(t-1)) .$$

Except for chapter 4 we assume that f is an odd function $(f(-x) = -f(x))$ and a periodic solution of (0.1) is called a special periodic solution provided

$$(0.2) \qquad \begin{cases} x(t+2) = -x(t) & \text{and} \\ x(-t) = -x(t) & \text{for all } t . \end{cases}$$

In general the phase space of (0.1) is typically infinite dimensional and the possible dynamics of solutions even for very simple looking f's is very complicated. By restricting to solutions of (0.1) which satisfy (0.2) the degrees of freedom in (0.1) are drastically reduced. For non-linearities f which satisfy $x \cdot f(x) > 0$, whenever $x \neq 0$, Kaplan and Yorke [19] have observed that any periodic solution of period exactly 4 of the cyclic system of ODE's rotating counterclockwise around the origin

$$(0.3) \qquad \begin{cases} \dot{x}(t) = -\alpha f(y(t)) \\ \dot{y}(t) = \alpha f(x(t)) \end{cases}$$

gives rise to a special periodic solution $x(t)$ of (0.1). Thus, by restricting attention to special periodic solutions, the infinite-dimensional phase space problem reduces to a 2-dimensional problem of finding a periodic solution of exactly period 4. Obviously, if $\{x(t), y(t)\}$ is a periodic solution of (0.3) of period T, then by a simple change of time scale one obtains a periodic solution of (0.3) of period 4 for a new parameter

$$\tilde{\alpha} = \alpha \cdot {}^{T}/_{4}$$

and therefore it suffices to study periodic solutions of (0.3) and their respective periods.

There is another way of studying solutions which satisfy (0.2) for (0.1). This is by putting the problem into a suitable operator form and that is

$$(0.4)\quad \begin{cases} \alpha Fx = x \\ \text{where} \quad (Fx)(t) = \int_0^t f(x(s-1))ds \\ \text{and} \quad x \in X \text{ , and } X \text{ is the Banachspace} \\ \text{of all continuous functions satisfying (0.2) .} \end{cases}$$

This approach is suitable if one wants to look at the problem from the point of view of bifurcation theory or the theory of nonlinear eigenvalue problems.

Our gool is to understand the class of special periodic solutions of (0.1) with respect to two major questions

- existence of bifurcation and continua of special periodic solution as α varies; multiplicity results for α fixed,
- numerical approximations of special periodic solutions.

Specifically we will be interested in a special class of nonlinearities. In most of the literature of equation (0.1) one finds the restriction $f(x) \cdot x > 0$ (for $x \neq 0$) and with respect to bifurcation from zero and infinity the typical assumption that

$$(0.5)\quad \begin{cases} f(s)/_s \to m_0 & \text{as} \quad |s| \to 0 \quad \text{and} \\ f(s)/_s \to m_\infty & \text{as} \quad |s| \to \infty \text{ .} \end{cases}$$

We will allow f to have several zeros and instead of (0.5) we will only require

$$(0.6)\quad \begin{cases} f(s)/_s \to m_0 \quad \text{as} \quad |s| \to 0 \\ m_\infty^2 \geq f(s)/_s \geq m_\infty^1 \quad \text{as} \quad |s| \to \infty \text{ .} \end{cases}$$

Thus, f has no derivative at ∞ and since f changes sign also in points other than 0 one will observe that the phase plane studies for (0.2) become nontrivial. Also the by now more or less classical approach of putting the periodicity problem for (0.1) into the setting of cones and positive operators is not applicable.

From a numerical point of view one would consider both the phase plane approach given by (0.3) as well as the fixed point approach given by (0.4) suitable. Recently, it was observed that standard numerical approximations for nonlinear elliptic boundary value problems [6,33,35,45] may produce 'spurious solutions'. These are numerical solutions which by no means are approximations of the given problem. So far these spurious solutions have only been understood by a careful and simultaneous investigation of the given equation and its numerical approximations. As typical reprensentatives for popular discretizations we will discuss here

- an implicit Runge-Kutta method for (0.3)

- a standard Newton-Cotes approximation for (0.4) .

Section 1 is devoted to a careful phase plane study both for the time-continuous problem (0.3) as well as a time-discrete numerical approximation given by implicit Newton formulas. The time-discrete model will be given by an area preserving diffeomorphism T , which is desireable because translation along trajectories of (0.3) also is an area-preserving diffeomorphism. It will be shown that though (0.3) is a perfect integrable system the portrait of T amounts to a surface of section of a nonintegrable system. We will discuss its homoclinic and heteroclinic structure and will introduce the concept of '<u>homoclinic</u> <u>bifurcation</u>' which describes a new scenario of phase transitions in the stochastic regime of one-parameter families of area preserving diffeomorphisms. In summary the analysis of T will provide numerous types of spurious solutions for (0.3).

Section 2 is devoted to the study of (0.4) in view of the global existence of continua of special periodic solutions and their bifurcation behavior. The main result is Theorem 2.1. Its proof is a blend of careful a priori estimates for special periodic solutions and a global perturbation technique for nonlinear eigenvalue problems which exploits the Global Bifurcation Theorem of Rabinowitz and yields bifurcation from infinity without assuming (0.5) (see[34,36]).

Section 3 again is a numerical study which is based on the trapezoidal rule approximation for (0.4). The main result is Theorem 3.3 which provides spurious solutions, the number of which grows like k^n , where k is the number of positive zeros of f and $n \in N$ is the number of meshpoints in the composite trapezoidal rule.

Section 4 is a theoretical study of (0.1) again. All previous sections strictly depend on the assumption of <u>oddness</u> for f . Here we attempt to relax this condition. The approach is new in so far as we will blend

- phase plane ideas
  with

- abstract methods from nonlinear functional analysis.

The main results are Theorems 4.1, 4.2 and 4.3 which give existence theorems of periodic solutions of (0.1) for a new class of nonlinearities.

Acknowledgements

The work of the first author was partially supported by a National Science Foundation grant and the work of the second author was partially supported by a grant from Stiftung Volkswagenwerk.

CHAPTER 1

1. PHASE PLANE STUDIES OF A CYCLIC SYSTEM

In this section we shall study time continuous- and time-discrete phase planes associated with the cyclic system

$$(1.1)\qquad \begin{cases} \dot{x}(t) = - f(y(t)) \\ \dot{y}(t) = f(x(t)) , \end{cases}$$

where $f: R \to R$ is an odd (i.e. $f(-x) = - f(x)$) and locally Lipschitzian function satisfying

$$(1.2)\qquad \begin{cases} \{z \geq 0 : f(z) = 0\} = \{0,z_1,z_2\} \quad \text{and} \\ 0,z_1 \text{ and } z_2 \text{ are simple,} \\ f(s) > 0 , \text{ for all } s \in (0,z_1) \text{ and } s \in (z_2,\infty) \\ f(s) < 0 , \text{ for all } s \in (z_1,z_2) \end{cases}$$

Typically f is given as in figure 1:

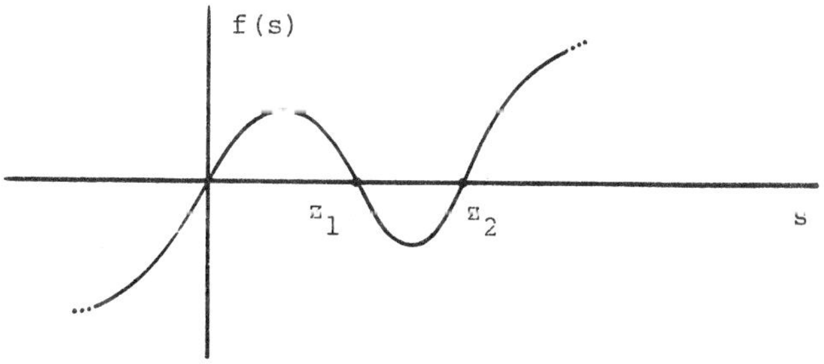

Figure 1.

The results following can be easily extended to more general nonlinearities, e.g. oscillating functions with more than two positive zeroes. We restrict to a discussion of example (1.2) only for reasons of length and technical complication. Let

$H: R^2 \to R$ denote the map

$$H(x,y) = \int_0^x f(s)ds + \int_0^y f(s)ds$$

then (1.1) can equivalently by written in the form

$$(1.3)\qquad \begin{cases} \dot{x} = - H_y \\ \dot{y} = H_x \end{cases}$$

Received by the editors March 10, 1983.

(H_y (resp. H_x) denoting the partial with respect to y (resp. x)) . Thus, (1.1) is an elementary example of an integrable Hamiltonian system with the constant of motion

$$H(x(t),y(t)) \equiv \text{constant} ,$$

and to describe the dynamics of (1.1) in the (x,y)-plane amounts to a discussion of the level curves of $H \equiv c$, which are smooth manifolds of dimension 1 , whenever $c \in R$ is a regular value. Observe that a point $c \in R$ is a regular value for H provided

$$H^{-1}(c) \cap F = \phi$$

where F is the set of rest points of (1.1), i.e.

$$F = \{(x,y) : x,y \in f^{-1}(0)\} .$$

Let $\Phi^t : R^2 \to R^2$, $t \in R$, denote the <u>phase flow</u> given by the vectorfield

$$V(x,y) = (-f(y),f(x)) ,$$

i.e. $\frac{d}{dt}(\Phi^t(\cdot)) = V(\Phi^t(\cdot))$. To study particular flow lines of Φ^t it will be useful to observe that the phase portrait is invariant under the action of the group G of order 8 generated by the involutions

$$(1.4) \quad \begin{cases} a(x,y) = (x,-y) & \text{(reflection in the x-axis)} \\ b(x,y) = (y,x) & \text{(reflection in } \Delta = \{(x,y) : x=y\}) \end{cases}$$

and the defining relations

$$a^2 = b^2 = (ab)^4 = (ba)^4 = Id$$

Note that one of the 3 invariant subgroups of order 4 is $D_4 \subset G$, the group of rotations by 90° , which is generated by ba . Now taking into account the oddness of f one has for the vectorfield V :

$$(1.5) \quad \begin{cases} \alpha^{-1}V\alpha = -V & \text{for } \alpha \in S \\ \alpha^{-1}V\alpha = V & \text{for } \alpha \in D_4 \end{cases}$$

where $S = \{a,b,aba,bab\}$

Let $P \in F$ be a saddle for Φ^t . As usual we denote by

$$W^s(P) = \{(x,y) : \Phi^t(x,y) \to P \quad \text{as} \quad t \to +\infty\}$$

the <u>stable manifold</u> of P

and by

$$W^u(P) = \{(x,y) : \Phi^t(x,y) \to P \quad \text{as} \quad t \to -\infty\}$$

the <u>unstable manifold</u> of P

Collecting the facts we have discussed so far we have (assuming for convenience that f is C^1):

<u>LEMMA 1.1.</u>

(1) For any $t \in R$ the map Φ^t is an area preserving diffeomorphism

(2) Let $P \in F$. Then $P = (x,y)$ is a
center provided $f'(x) \cdot f'(y) > 0$
saddle provided $f'(x) \cdot f'(y) < 0$.

(3) The phase portrait of Φ^t is invariant under G, i.e. more precisely

(i) $(\Phi^t\alpha)^2 = (\alpha\Phi^t)^2 = Id$ for all $t \in R$ and $\alpha \in S$.

(ii) $\Phi^t\alpha = \alpha\Phi^t$ for all $t \in R$ and $\alpha \in D_4$.

(iii) Define $R_t^\alpha := \Phi^t\alpha$, $t \in R$. Then $\Phi^t = R_t^\alpha R_o^\alpha$ for all $\alpha \in S \cup \{Id, (ab)^2\}$, i.e. the phase flow can be factored into involutions. (Note that $R_o^\alpha = \alpha$ and $(ab)^2 = -Id$.)

(iv) Let $P \in F$ be a saddle. Then

$$R_t^\alpha(W^u(P)) = W^s(\alpha(P))$$
$$R_t^\alpha(W^s(P)) = W^u(\alpha(P))$$

for $\alpha \in S$, and

$$R_t^\alpha(W^u(P)) = W^u(\alpha(P))$$
$$R_t^\alpha(W^s(P)) = W^s(\alpha(P))$$

for $\alpha \in D_4$.

<u>PROOF</u>. Observe that $div(V) = 0$ and therefore Liouville's theorem implies that Φ^t is area preserving. Thus, all restpoints are either centers or saddles and (2) is obvious. To prove (3) we use (1.5): Let $\alpha \in S$, i.e. $\alpha = \alpha^{-1}$. Then

$$\alpha V \alpha = - V \quad \text{implies that} \quad \alpha\Phi^t\alpha = \Phi^{-t}$$

and this implies (i). The assertion (ii) follows similarily and assertion (iii) is just a rewriting of (i) and (ii). To show (iv), let $Q \in W^s(P)$. Then one has to show that

$$R_t^\alpha(Q) \in W^u(\alpha(P)) \quad \text{for} \quad \alpha \in S$$

$$R_t^\alpha(Q) \in W^s(\alpha(P)) \quad \text{for} \quad \alpha \in D_4 \ .$$

Let $\alpha \in S$. Consider $\Phi^{-s}(R_t^\alpha(Q)) = \Phi^{-s}\,\Phi^t\alpha(Q)$. Now we have from (i) that $\Phi^{-s} = \alpha\Phi^s\alpha$, and, thus, $\Phi^{-s}\,\Phi^t\alpha(Q) = \alpha\Phi^{s-t}(Q) \to \alpha(P)$, as $s \to +\infty$, i.e. $R_t^\alpha(Q) \in W^u(\alpha(P))$. Finally, let $\alpha \in D_4$. Consider $\Phi^s(R_t^\alpha(Q)) = \Phi^s\,\Phi^t\alpha(Q)$. Now we use (ii) to conclude that $\Phi^s\,\Phi^t\alpha(Q) = \alpha\Phi^{s+t}(Q) \to \alpha(P)$, as $s \to +\infty$. So far we have proved that $R_t^\alpha(W^s(P)) \subset W^u(\alpha(P))$ (resp. $R_t^\alpha(W^s(P)) \subset W^s(\alpha(P))$. The desired identities follow from (i) and (ii)

As an immediate consequence we note that the above symmetries allow to reduce the study of the phase portrait of Φ^t to a study in C_{45} , the first 45° section of the first quadrant, i.e.

$$C_{45} = \{(x,y) : x \geq 0 , \quad y \geq 0 , \quad y \leq x\} \ .$$

Before we discuss the phase plane of (1.1) we introduce a special time-discrete dynamical system $T: \mathbb{R}^2 \to \mathbb{R}^2$: Discretizing (1.1) by an implicit Euler method one obtains the difference equations

$$(1.6) \qquad \begin{cases} x_{n+1} = x_n - hf(y_n + hf(x_n)) \\ y_{n+1} = y_n + hf(x_n) \ , \end{cases}$$

where $h > 0$ denotes the stepsize. Equations (1.6) induce the map $T = T_2 \circ T_1$, where

$$T_1(x,y) = (x,y + hf(x)) \text{ and } T_2(x,y) = (x-hf(y),y) \ .$$

Given a point $Q \in R$, then in an obvious sense the orbit $\bigcup_{n\in\mathbb{Z}} T^n(Q)$ provides a numerical approximation of the flow line of Φ^t through Q . In our definition $T = T(f,h)$, i.e. we have a model $\mathcal{T}$ of dynamical systems T , where we consider $f: R \to R$ and $h \in R$ as parameters. Mostly, we simply suppress the dependence on f and h , but sometimes we will stress it by writing T_h (resp. T_f) , meaning that we fix f and let h vary (resp. we fix h and let f vary).

The literature of numerical initial value solvers is very rich (see e.g. [7,12,17,18]). Very popular are

- one-step methods, such as Runge-Kutta
  and

- multi-step methods, such as Adams-Bashforth.

Our T reflects an implicit Runge-Kutta method. Of course in view of higher accuracy and efficiency one would prefer e.g. a higher order Runge-Kutta method. A discussion of such a method could be carried through by essentially the same arguments, we shall give below, but with a tremendous increase in technical complication. In any case, the choice of a numerical method would be governed by the desire to derive a scheme which inherits as much as possible of the structure which is intrinsic to (1.1). Typical intrinsic properties are collected in lemma 1.1. In view of these we investigate briefly the properties of T (again assuming for convenience that $f \in C^1$). Rest points of Φ^t will correspond to fixed points of T:

$$T(x,y) := (x-hf(y+hf(x)), \quad y+hf(x)) = (x,y)$$

Thus, $F = \{(x,y) : T(x,y) = (x,y)\}$. Note that for all $z = (x,y) \in R^2$ one has

$$T'(z) = T_2'(T_1(z)) \circ T_1'(z), \quad \text{with}$$

$$T_2'(\cdot) = \begin{pmatrix} 1 & -hf'(y) \\ 0 & 1 \end{pmatrix}, \quad T_1'(\cdot) = \begin{pmatrix} 1 & 0 \\ hf'(x) & 1 \end{pmatrix}.$$

Thus, $\det(T'(z)) = 1$, for all $z \in R^2$. As usual we distinguish the following types of fixed points $P \in F$ by considering the eigenvalues λ_1, λ_2 of $T'(P)$:

$$(1.7) \quad \begin{cases} P \text{ elliptic} & \Leftrightarrow \lambda_1,\lambda_2 \in S^1\setminus\{(1,0),(-1,0)\} \subset C \\ P \text{ hyperbolic} & \Leftrightarrow 0 < \lambda_1 < 1, \quad \lambda_2 > 1 \\ P \text{ inverse hyperbolic} & \Leftrightarrow -1 < \lambda_1 < 0, \quad \lambda_2 < -1 \\ P \text{ parabolic} & \Leftrightarrow |\lambda_1| = |\lambda_2| = 1 \text{ and } \lambda_j \text{ real for } j=1,2. \end{cases}$$

We also recall the following definition:

DEFINITION 1.1. ([25,38]) Let $P \in F$ be a hyperbolic fixed point of T, then

$W^s(P) = \{z : T^n(z) \to P \text{ as } n \to \infty, n \in Z\}$
is called the stable manifold of P, and

$W^u(P) = \{z : T^n(z) \to P \text{ as } n \to -\infty, n \in Z\}$
is called the unstable manifold of P.

If f is of class C^1 then $W^s(P)$ and $W^u(P)$ are immersed smooth 1-manifolds (cf. [38]).

DEFINITION 1.2. (cf. [25,38]) Let $P_1, P_2 \in F$ be hyperbolic fixed points of T.

(i) $Q \in W^s(P_1) \cap W^u(P_1) \setminus \{P_1\}$ is called a transversal (resp. degenerate) homoclinic point, provided $W^s(P_1)$ and $W^u(P_1)$ have a transversal (resp. tangential) intersection in Q.

(ii) $Q \in W^s(P_1) \cap W^u(P_2) \setminus \{P_1, P_2\}$ is called a transversal (resp. degenerate) heteroclinic point, provided $W^s(P_1)$ and $W^u(P_2)$ have a transversal (resp. tangential) intersection in Q.

Recall the famous Birkhoff-Smale theorem [4,38] saying, that given a transversal homoclinic point Q for T there exists an $N > 0$ and a set Λ invariant under T^N containing Q, such that

- Λ contains a dense set of periodic points
- T^N is mixing on Λ
- Λ contains a dense T^N-orbit.

Since $\det(T'(z)) = +1$ for all $z \in R^2$, it is obvious that (1.7) covers all possible cases.

In lemma 1.1 we have emphasized the important role played by the group G of symmetries of Φ^t. The question arises whether the time-discrete phase portrait of T has analogous symmetries. A first observation is that G does not leave the phase portrait of T invariant, e.g. $(Ta)^2 \neq Id$. We define the following involutions of R^2 for $h \in R$ and any odd function $f: R \to R$:

$$(1.8)\quad \begin{cases} a_h(x,y) = (x+hf(y), -y) & \\ b(x,y) = (y,x) & ; b_h \equiv b \\ c_h(x,y) = (-x, y+hf(x)) & \\ d(x,y) = (-y, -x) & ; d_h \equiv d \end{cases}$$

Note that $a_o = a$, and $c_h = ba_hb$, and $d = aba$. However, one has $d \neq a_h b a_h$ for $h \neq 0$ in general. Now let G_h denote the group generated by the involutions of (1.8). Then a first remarkable difference to the continuous-time dynamical system Φ^t is that G_h is in general not a finite group for $h \neq 0$, though $G_o = G$. In G we have the relation $(ab)^2 = (ba)^2 = -\,Id$. Correspondingly, in G_h we compute that

$$(1.9)\qquad \begin{cases} a_h b = bc_h = -\,a_h d = -\,dc_h \\ c_h d = da_h = -\,ba_h = -\,c_h b \\ (a_h b)^2 = -\,T_h;\ (ba_h)^2 = -\,T_h^{-1} . \end{cases}$$

Thus, T_h^{-1} exists for all $h \in R$, and $T_h = Id$ for $h = 0$, and $-\,T_h$ and $-\,T_h^{-1}$ have a square root. Thus, if we wish to compare the continuous-time and the discrete-time systems we see that the correspondant of $D_4 \subset G$ is the infinite group H generated by $\sqrt{-T_h}$. Of course there are many other ways of comparing G_o and G_h, e.g. for any $z \in R^2 \setminus \{0\}$ the isotropy group $D_4^z = \{g \in D_4 : g(z)=z\} = \{Id\}$, whereas for $h \neq 0$ H^z may be nontrivial, e.g. if $T^p(z) = z$ for some $p \in N$. In view of lemma 1.1 we wish to concentrate on common properties of Φ^t and T:

<u>LEMMA 1.2</u>. Let f be C^1. Then

(1) $T(f,h)$ is an area preserving diffeomorphism.

(2) Let $P \in F$ and $P = (x,y)$. Set

$$\tau_h(x,y) = h^2 f'(x) f'(y) .$$

Then P is

elliptic, provided	$0 < \tau_h < 4$
hyperbolic, provided	$\tau_h < 0$
inverse hyperbolic, provided	$\tau_h > 4$
parabolic, provided	$\tau_h = 0\ ,\ 4$.

(3) The phase portrait of T has the following symmetries:

(i) $(T^n \alpha_h)^2 = (\alpha_h T^n)^2 = Id$ for all $n \in Z$

and $\alpha_h \in S_h := \{a_h, b_h, c_h, d_h\}$.

(ii) $T^n \alpha_h = \alpha_h T^n$ for all $n \in Z$ and

$\alpha_h \in D_h := \{a_h b, a_h d, c_h d, c_h b\}$

(iii) Define $R_n^{\alpha_h} := T^n \alpha_h$, $n \in Z$. Then

$$T^n = R_n^{\alpha_h} R_o^{\alpha_h}$$

for all $\alpha_h \in S_h$, i.e. the discrete-time dynamical system can be factored into involutions. (Note that $R_o^{\alpha_h} = \alpha_h$.)

(iv) Let $P \in F$ be a hyperbolic fixed point. Then

$$R_n^{\alpha_h}(W^u(P)) = W^s(\alpha_o(P))$$

$$R_n^{\alpha_h}(W^s(P)) = W^u(\alpha_o(P))$$

for $\alpha_h \in S_h$, $n \in Z$, and

$$R_n^{\alpha_h}(W^u(P)) = W^u(\alpha_o(P)) ,$$

$$R_n^{\alpha_h}(W^s(P)) = W^s(\alpha_o(P))$$

for $\alpha_h \in D_h$, $n \in Z$.

PROOF. We have seen already that T^{-1} exists and that $\det(T'(\cdot))=+1$. This proves (1). To see (2) we compute the eigenvalues λ_1, λ_2 of $T'(x,y)$, where $P = (x,y) \in F$, to be the roots of

$$\lambda^2 - \lambda \text{ trace } (T'(x,y)) + 1 = 0 ,$$
with trace $(T'(x,y)) = 2 - \tau_h$.

Thus, $$\lambda_{1,2} = \frac{2-\tau_h \pm \sqrt{\tau_h^2 - 4\,\tau_h}}{2} .$$

To show (3) we make use of (1.9): Let $\alpha_h \in S_h$. Then we choose an appropriate representation for T according to (1.9), i.e.

$$T = -(\alpha_h\beta_h)^2 \quad \text{or} \quad T = -(\beta_h\alpha_h)^2 .$$

Then $$(T^n\alpha_h)^2 = \underbrace{\alpha_h\beta_h \cdots \alpha_h\beta_h}_{2n\text{-times}} \cdot \alpha_h \cdot \underbrace{\alpha_h\beta_h \cdots \alpha_h\beta_h}_{2n\text{-times}} \cdot \alpha_h$$

$$= \mathrm{Id} .$$

The computation for the second representation is analogous. This proves (i). To prove (ii) one observes that $T^n\alpha_h = \alpha_h T^n \Leftrightarrow \alpha_h T^{-n} = T^{-n}\alpha_h$ and that for any choice of $\alpha_h \in D_h$ there exists an appropriate representation in (1.9), i.e. either $(-T)^n = (\alpha_h)^{2n}$ or $(-T)^{-n} = (\alpha_h)^{2n}$, and in either case, (ii) is then obvious. Assertion (iii) is just a rewriting of (i). Finally to show (iv) we carry out an induction via $n \in N$:
n = 0: Let $z \in W^s(P)$. We show that

$$\alpha_h(z) \in W^u(\alpha_0(P)) \quad \text{for} \quad \alpha_h \in S_h \ , \quad \text{and}$$

$$\alpha_h(z) \in W^s(\alpha_0(P)) \quad \text{for} \quad \alpha_h \in D_h \ .$$

In the first case, we have $T^k(z) \to P$ as $k \to +\infty$. Thus, $\alpha_h T^k(z) \to \alpha_h(P) = \alpha_0(P)$. But according to (i) $\alpha_h T^k(z) = \alpha_h T^k \alpha_h \alpha_h(z) = (\alpha_h T \alpha_h)^k \alpha_h(z) = T^{-k}\alpha_h(z)$, and, hence, $\alpha_h(z) \in W^u(\alpha_0(P))$. In the second case, we have $\alpha_h T^k(z) \to \alpha_0(P)$ as $k \to +\infty$. According to (ii) $\alpha_h T^k = T^k \alpha_h$, i.e. $\alpha_h(z) \in W^s(\alpha_0(P))$.

<u>$n-1 \to n$:</u> Let $z \in W^s(P)$. We have that

$$T^{n-1}\alpha_h(z) \in W^u(\alpha_0(P)) \quad \text{for} \quad \alpha_h \in S_h \ , \quad \text{and}$$

$$T^{n-1}\alpha_h(z) \in W^s(\alpha_0(P)) \quad \text{for} \quad \alpha_h \in D_h \ .$$

In the first case, consider $T^{-k}(T^n\alpha_h(z)) = T^{-k+1}(T^{n-1}\alpha_h(z)) \to \alpha_0(P)$, as $k \to +\infty$, i.e. $R_n^{\alpha_h}(z) \in W^u(\alpha_0(P))$. In the second case, consider $T^k(T^n\alpha_h(z)) = T^{k+1}(T^{n-1}\alpha_h(z)) \to \alpha_0(P)$, as $k \to +\infty$, i.e. $R_n^{\alpha_h}(z) \in W^s(\alpha_0(P))$. So far we have proved that $R_n^{\alpha_h}(W^s(P)) \subset W^u(\alpha_0(P))$ for $\alpha_h \in S_h$ and $R_n^{\alpha_h}(W^s(P)) \subset W^s(\alpha_0(P))$ for $\alpha_h \in D_h$. The other cases are proved analogously. Finally, to obtain the desired identities we use (i) and (ii) again:
$R_n^{\alpha_h}(W^s(P)) \subset W^u(\alpha_0(P))$ and $R_n^{\alpha_h}(W^u(P)) \subset W^s(\alpha_0(P))$ implies $(R_n^{\alpha_h})^2 W^s(P) \subset R_n^{\alpha_h}(W^u(\alpha_0(P))) \subset W^s(P)$, i.e. $R_n^{\alpha_h}(W^u(\alpha_0(P))) = W^s(P)$ and substituting P by $\alpha_0(P)$ gives $R_n^{\alpha_h}(W^u(P)) = W^s(\alpha_0(P))$. In the other case $R_n^{\alpha_h}(W^s(P)) \subset W^s(\alpha_0(P))$ implies $W^s(P) \subset (R_n^{\alpha_h})^{-1}W^s(\alpha_0(P)) \subset W^s(P)$, i.e. $(R_n^{\alpha_h})^{-1}W^s(\alpha_0(P)) = W^s(P)$ or equivalently $R_n^{\alpha_h}(W^s(P)) = W^s(\alpha_0(P))$.

An immediate consequence is

<u>REMARK 1.1</u>. Let $P = (x,y) \in F$ be a fixed point of T_h with $\tau_h = h^2 f'(x)f'(y) > 0$. Set

$$h_1 = 2\sqrt{\{f'(x)f'(y)\}^{-1}}$$

Then $h = h_1$ is a point of bifurcation for the one-parameter family T_h^2. More precisely, as h passes through h_1 from below the elliptic fixed point P of T_h becomes inverse hyperbolic for T_h.

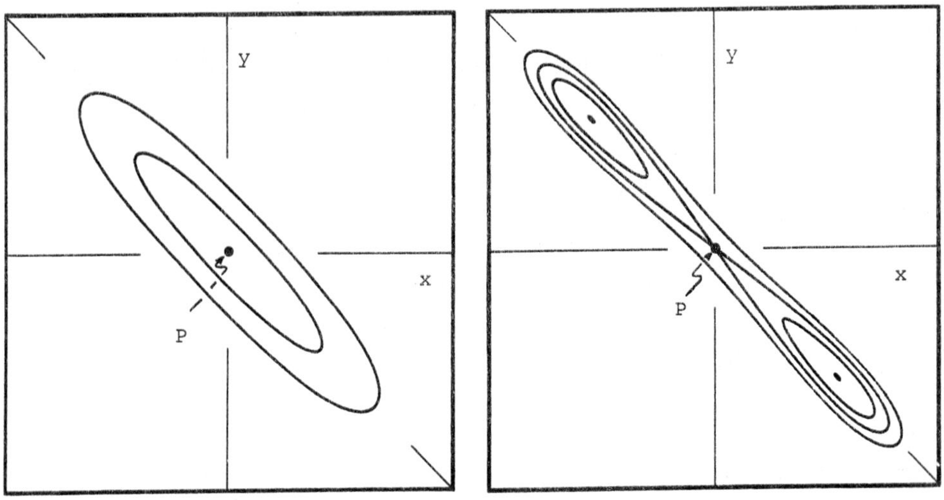

Figure 2.

Figure 2 shows a computer experiment for $f(s) = \sin(s)$, $P = (0,0)$, $h_1 = 2$, $\underline{h} = 1.9$, $\bar{h} = 2.1$. The window shown is $\square = \{(x,y) : -2.5 \le x,y \le 2.5\}$. Here the bifurcation is into 2 elliptic fixed points E_1 , E_2 of T_h^2 with

$$\begin{cases} T_h(E_1) = E_2 \ , \quad T_h(E_2) = E_1 \quad \text{and} \\ E_1, E_2 \in \{(x,y) : x = -y\} = a_o(\Delta) \end{cases}$$

We observe that $d(a_o(\Delta)) = a_o(\Delta)$. Due to the symmetries (1.8) and the relations (1.9) we have that for any fixed point z of T_h^2 also $\alpha_h(z)$ is a fixed point of T_h^2 for $\alpha_h \in \{a_h, b_h, c_h, d_h\}$. This fact provides a possibility to compute E_1 and E_2 explicitly, when we assume that E_1, E_2 are the only fixed points of T_h^2 which bifurcate from $P = (0,0)$. Then we must have that

$$\alpha_h(\{E_1,E_2\}) = \{E_1,E_2\} \quad ,$$

which yields the equations

$$\begin{cases} 2x = -hf(y) \\ 2y = -hf(x) \quad , \end{cases}$$

and, hence,

$$x = \frac{h}{2}f(\frac{h}{2}f(x)) ,$$

which is solvable for $f(s) = \sin(s)$ and $h = 2 + \varepsilon$. Thus, the discussion of the two-dimensional fixed point problem may be reduced to a scalar problem by use of the symmetries. To see that h_1 is indeed a point of bifurcation, we use a known argument from general bifurcation theory. Let $i_{R^2}(T_h^2,U)$ denote the fixed point index of T_h^2 in a neighborhood U of P , which isolates P for $h = h_1 - \varepsilon$ and $h_1 + \varepsilon$, ε small. Then

$$i_{R^2}(T_h^2,U) = +1 , \quad h = h_1 - \varepsilon$$
$$i_{R^2}(T_h^2,U) = -1 , \quad h = h_1 + \varepsilon ,$$

because for $h = h_1 + \varepsilon$ precisely one eigenvalue of $T_h'(P)$ is less than -1 , and therefore precisely one eigenvalue of $(T_h^2)'(P) = \{T_h'(P)\}^2$ is greater than $+1$. The change of index implies bifurcation.

Once, one has observed the bifurcation at h_1 one may guess that this mechanism may repeat: The two elliptic fixed points of T_h^2 become unstable at $h = h_2$ and give birth to four elliptic fixed points of T_h^4 , etc. One thus would obtain a sequence of critical parameters $h_1, h_2, h_3, \ldots$ at which period doubling bifurcation occurs. Such a mechanism would be analogous to the celebrated study of Feigenbaum for the 1-dimensional quadratic map $x \to 1-\mu x^2$, $\mu > 0$, which shows period doubling behaviour at critical parameters $\mu_1, \mu_2, \mu_3, \ldots$ with $\mu_k \to \mu_\infty < \infty$ and

$$\frac{\mu_k - \mu_{k-1}}{\mu_{k+1} - \mu_k} \to \delta \text{ as } k \to \infty ,$$

with $\delta = 4.66920\ldots$ Feigenbaum's remarkable observation was that δ is universal in the sense that it characterizes a whole class of nonlinearities f close to a quadratic map (cf. [8,11,15]). In analogy one would expect that the h_k show some universal behaviour with respect to their asymptotic ratios. Indeed recent research (cf. [9,14]) seems to indicate that this is true.

If we compare lemma 1.1 with lemma 1.2 we observe that many fundamental properties of the continuous-time dynamical system Φ^t are inherited by the discrete-time dynamical system T . We thus might conjecture that at least for h small the phase portraits should be approximately the same. The main goal of this section is to indicate that this is not the case: Φ^t is the flow of an integrable system, whereas T has all the typical properties of a Poincaré-map of a non-integrable two degree of freedom Hamiltonian system (cf. [14,16]). For special choices of nonlinearities f we will show that the homoclinic and heteroclinic orbits of Φ^t will split when passing to $T(f,h)$. These homo- and heteroclinic structures are responsible for a stochastic dynamic of T near hyperbolic fixed points. It seems to be noteworthy that higher order Runge-Kutta methods lead basically to the same phenomena. In our example the symmetries $R_n^{\alpha h}$ will be of fundamental importance to understand the homo- and heteroclinic structures. The idea of using symmetries to investigate the dynamics of area preserving diffeomorphisms goes back to G.D. Birkhoff [5] and R. De Vogelaere [43] (see also J.M. Greene [13]), who used involutions to study periodic points. Our approach here is a continuation of recent results in [31,32], where symmetries like $R_n^{\alpha_h}$ played a key role in understanding the interplay of homoclinic and periodicity structures of one-parameter families of area preserving diffeomorphisms.

We now will discuss the continuous-time phase planes of (1.1) for typical choices of nonlinearities f. Since it is basically known and elementary, how to construct the phase portraits of such systems from the constant of motion H (cf. [3]) we restrict ourselves to a sketchy discussion revealing only the essential steps. We also restrict to a nonlinearity as given in figure 1 and assume for convenience that $f \in C^1$ and that all zeroes are simple. We will distinguish three qualitatively different situations:

(1.10) $$\int_0^{z_2} f(s)ds > 0$$

(1.11) $$\int_0^{z_2} f(s)ds = 0$$

(1.12) $$\int_0^{z_2} f(s)ds < 0$$

Case (1.10): We claim that the phase portrait of (1.1) restricted to C_{45} is given by figure 3, and figure 4 shows a computer plot of the whole phase portrait. The little arrows indicate the direction of the vectorfield

$$V(x,y) = (-f(y),f(x)) .$$

To verify figure 3 we set $F(z) = \int_0^z f(s)ds$ and $P_o = (0,0)$, $P_1 = (z_1,0)$, $P_2 = (z_2,0)$, $P_3 = (z_1,z_1)$, $P_4 = (z_2,z_1)$, $P_5 = (z_2,z_2)$.

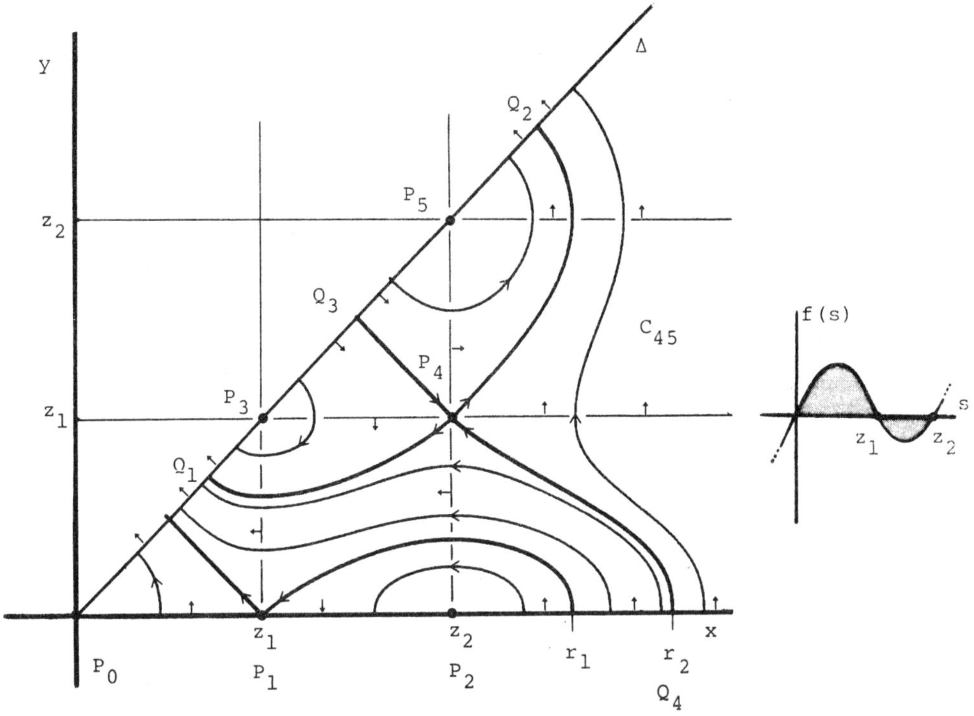

Figure 3.

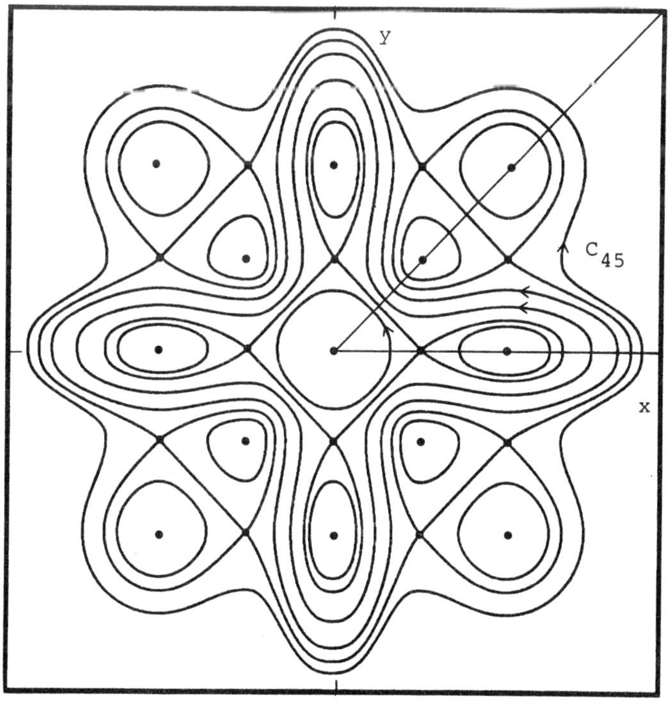

Figure 4.

It suffices to characterize the separatrices emanating from P_1 and P_4. The elliptic (periodic) motions around P_o, P_2, P_3 and P_5 are then a consequence of the symmetry property of the phase flow.

We begin to study the level curves of energy $H(z_2,z_1)$. By linearization of the vector field in P_4 one obtains that the stable manifold of P_4 approaches P_4 with slope $-\sqrt{-\beta/\alpha}$ and the unstable manifold of P_4 approaches P_4 with slope $\sqrt{-\beta/\alpha}$, where $\alpha = f'(z_1) < 0$ and $\beta = f'(z_2) > 0$. Thus, taking into account the direction of the vectorfield at the lines $x \equiv z_2$ and $y \equiv z_1$ one cannot have a homoclinic orbit emanating from P_4 <u>and</u> entirely lying in C_{45}. It is also obvious that the level curves emanating from P_4 cannot go to infinity in C_{45} in forward or backward time, because then $H(x,y) \to \infty$ as $x \to \infty$ or $y \to \infty$, but H is a constant of motion. Finally, a level curve emanating from P_4 cannot contain a point $(x,0)$ with $0 \le x \le z_2$, because otherwise

$$F(x) = F(z_2) + F(z_1) \Leftrightarrow F(x) - F(z_1) = F(z_2),$$

but $F(z_2) > 0$ by (1.10) and $F(x) - F(z_1) \le 0$. Thus, taking into account the direction of the vectorfield along the x-axis and the diagonal Δ we may conclude that these level curves which emanate from P_4 in forward time intersect the diagonal Δ transversally in two uniquely determined points

$$(1.13)\quad \begin{cases} Q_1 = (x_+^1, x_+^1) \text{ with } 0 < x_+^1 < z_1 \text{ and} \\ \qquad 2\int_o^{x_+^1} f(s)ds = F(z_1) + F(z_2) \\ \text{and} \\ Q_2 = (x_+^2, x_+^2) \text{ with } x_+^2 > z_2 \text{ and} \\ \qquad \int_{z_2}^{x_+^2} f(s)ds = -\frac{1}{2}\int_{z_1}^{z_2} f(s)ds \end{cases}$$

Similarily, in backward time we will have two level curves emanating from P_4, one intersecting the diagonal Δ transversally in the unique point

$$(1.14)\quad \begin{cases} Q_3 = (x_-^1, x_-^1) \text{ with } z_1 < x_-^1 < z_2 \text{ and} \\ \qquad \int_z^{x_-^1} f(s) = \frac{1}{2}\int_{z_1}^{z_2} f(s)ds \end{cases}$$

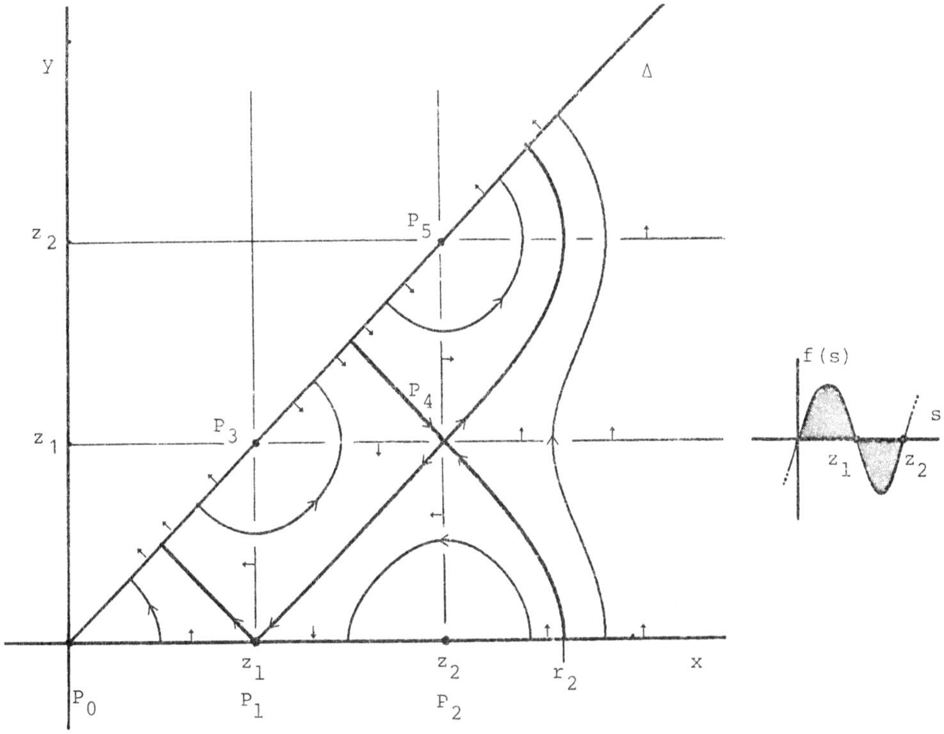

Figure 5.

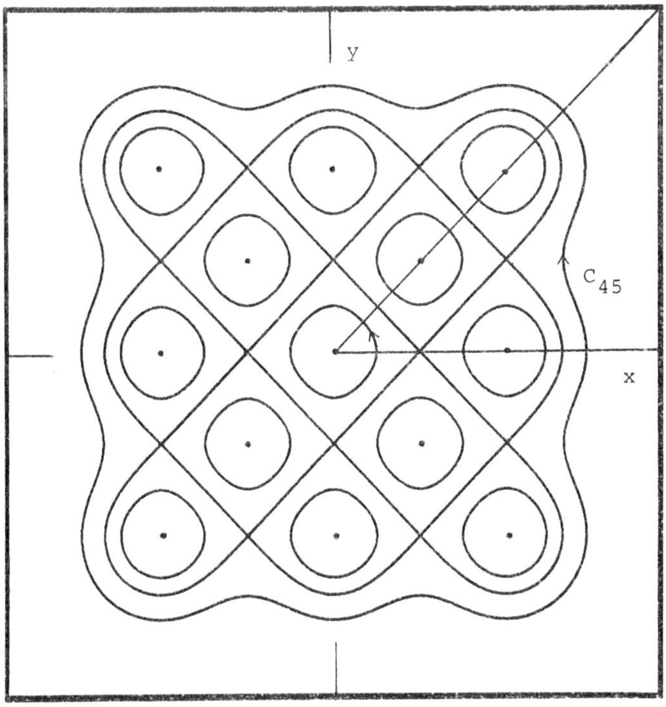

Figure 6.

and the other intersecting the x-axis transversally in the unique point

$$(1.15)\quad \begin{cases} Q_4 = (x_-^2, 0) \quad \text{with } x_-^2 > z_2 \text{ and} \\ \int_{z_2}^{x_-^2} f(s)ds = \int_0^{z_1} f(s)ds . \end{cases}$$

Note that the first equation in (1.13) is soluble if

$$\int_0^{z_1} f(s)ds + {}^1/_2 \int_{z_1}^{z_2} f(s)ds > 0 .$$

Condition (1.10) enters as follows: Let (x,y) be the (unique) point on the level curve containing P_4 and Q_1 with $x = z_1$ and $z_1 > y > 0$. Then $F(z_1) + F(y) = F(z_1) + F(z_2)$, i.e. $F(z_2) > 0$. In a very similar manner one can analyze the two (when restricting to C_{45}) level curves emanating from P_1 and verify figure 3. Note that though P_1 has no homoclinic orbit, which entirely lies in C_{45}, it has a homoclinic orbit: by the symmetry with respect to the x-axis, the level curve through $P_1 = (z_1, 0)$ and $(r_1, 0)$ constitutes a homoclinic orbit for (1.1). The other separatrices of C_{45} constitute heteroclinic orbits for (1.1). The points of intersection of the above separatrices with the x-axis are crucial for our discussion of spurious solutions in section 3. We therefore define:

$$(1.16)\quad \begin{cases} r_1 > z_2 \text{ by the condition } \int_{z_1}^{r_1} f(s)ds = 0 \\ \text{and} \\ r_2 > z_2 \text{ by the condition } \int_{z_2}^{r_2} f(s)ds = \int_0^{z_1} f(s)ds \end{cases}$$

Assumption (1.10) implies that $r_2 > r_1$.

Case (1.11): In a completely analogous analysis one obtains the phase portrait of (1.1) restricted to C_{45} as given by figure 5, and figure 6 shows a computer plot of the whole phase portrait. The remarkable change from (1.10) to (1.11) is expressed in the existence of the heteroclinic orbit from P_1 to P_4 for (1.11), which is due to the fact that $H(z_1, 0) = H(z_2, z_1)$. The point of intersection of the stable manifold of P_4 with the x-axis determines a crucial number r_2 as in (1.16).

Case (1.12): Figure 7 shows the phase portrait of (1.1) restricted to C_{45} and figure 8 shows a computer plot of the whole phase portrait. The computer plots were obtained by solving intial value problems given

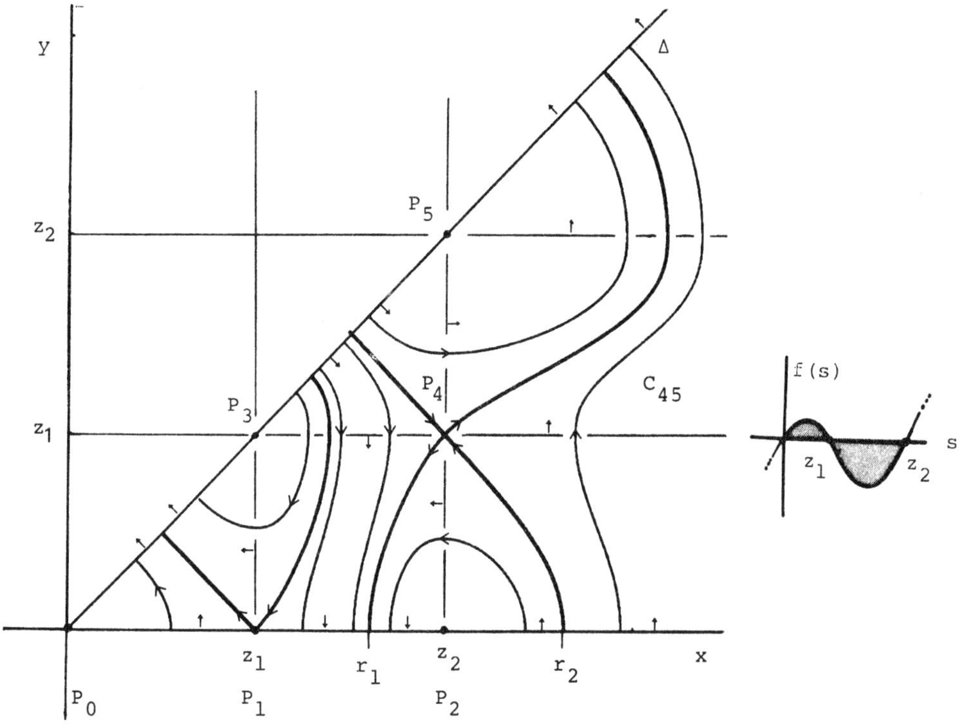

Figure 7.

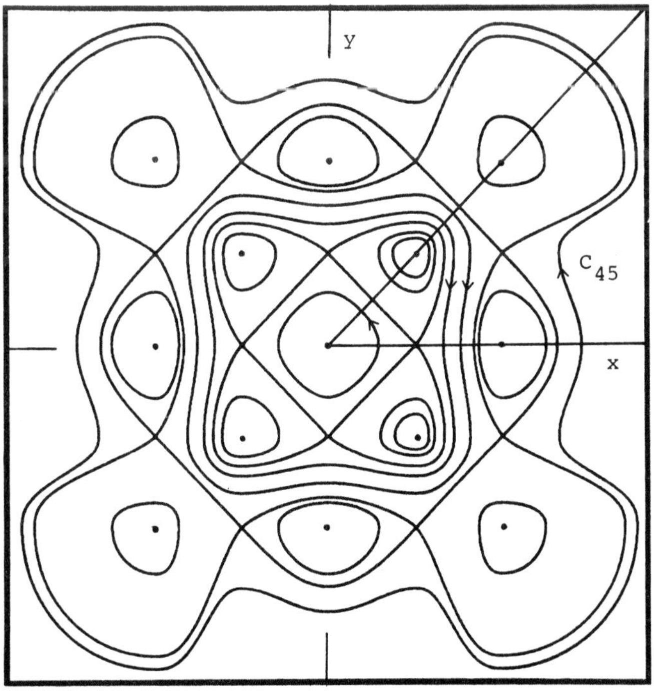

Figure 8.

by T with $h = 0.01$. In the discussion of T we have claimed that T will have splitting separatrices for all $h > 0$. This is indeed true, however this phenomenon is for $h = 0.01$ below visibility in the scale shown in figures 4, 6 and 8. Special experimental studies at the end of this section will provide evidence for this. Again we determine the points of intersection of the stable and unstable manifold of P_4 with the x-axis:

$$(1.17)\quad \begin{cases} z_1 < r_1 < z_2 \text{ by the condition } \int_{z_1}^{r_1} f(s)ds = \int_0^{z_2} f(s)ds \\ \text{and} \\ r_2 > z_2 \text{ by the condition } \int_{z_2}^{r_2} f(s)ds = \int_0^{z_1} f(s)ds . \end{cases}$$

We collect same useful observations:

REMARK 1.2

(1) Arbitrarily small perturbations of f can be used to switch from (1.11) into (1.10) or (1.12). Thus, system (1.1) is not structurally stable.

(2) In this paper we shall be interested in the study of periodic solutions of

$$\dot{x}(t) = -\alpha f(x(t-1)), \quad \alpha > 0$$

and in particular in special "special periodic solutions" such that $x(t+2) = -x(t)$ and $x(-t) = -x(t)$ for all t and $x(t) > 0$ for $0 < t < 2$. We shall see in the next section that such periodic solutions are (after allowing for time translation) in one-one correspondence with periodic solutions $(x(t), y(t))$ of

$$(1.18)\quad \begin{cases} \dot{x}(t) = -\alpha f(y(t)) \\ \dot{y}(t) = \alpha f(x(t)) \end{cases}$$

which have minimal period 4 and rotate counterclockwise about the origin. However, if $(\tilde{x}(t), \tilde{y}(t))$ is a periodic solution of (1.1) which rotates counterclockwise about the origin and has minimal period T, then $x(t) \equiv \tilde{x}(\frac{Tt}{4})$ and $y(t) \equiv \tilde{y}(\frac{Tt}{4})$ solves (1.18) for $\alpha = \frac{T}{4}$ and has minimal period 4. We have already analyzed the values $c > 0$ such that if $\tilde{x}(1) = c$ and $\tilde{y}(1) = 0$ and $(\tilde{x}(t), \tilde{y}(t))$ satisfies (1.1), then $(\tilde{x}(t), \tilde{y}(t))$ rotates counterclockwise about the origin until it hits the diagonal; the symmetries (1.5) then insure that $(\tilde{x}(t), \tilde{y}(t))$ is periodic. If, for such c, $\tau(c)$ denotes the first positive time τ such that $\tilde{x}(1+\tau) = 0$, the above remarks show that the differential-delay equation

$$\dot{x}(t) = -\alpha f(x(t-1))$$

has precisely as many special periodic solutions as the equation

$$\tau(c) = \alpha$$

has distinct positive solutions c . Thus a precise picture of the set of special periodic solutions for varying $\alpha > 0$ requires a precise understanding of the time map

$$c \to \tau(c) \; .$$

Such an understanding is known to be difficult even for quite special second order systems, as Smoller and Wassermann [39] have shown. The analysis of chapter 2 can be considered a study of $\tau(c)$ for c large and a very general class of f . However, if we confine ourselves to a rough qualitative picture here, we can argue that choosing initial values close to the intersections of the separatrices with the x-axis will produce periods $T >> 1$ and this implies after rescaling a periodic solution of period 4 of (1.18) with $\alpha >> 1$. Figures 9, 10 and 11 only show periodic solutions of (1.18) of period $T = 4$ for $\alpha >> 1$, which rotate counterclockwise around the origin. The diagrams show α versus x , where $(x,0)$ is the initial value determining the solution. Fat lines indicate continua of periodic solutions (parametrized over x) . The crucial property of periodic solutions of (1.1) which gives rise to solutions of the differential-delay equation is $v(t) = u(t-{}^{T}/_{4})$. This property may be satisfied for solutions rotating counterclockwise around other centers than $P_o = (0,0)$. For example, if $f(z_2+v)=:g(v)$ is odd for $|v|$ small, one obtains periodic solutions of the differential-delay equation. These solutions have period 4 and oscillate about the value z_2.

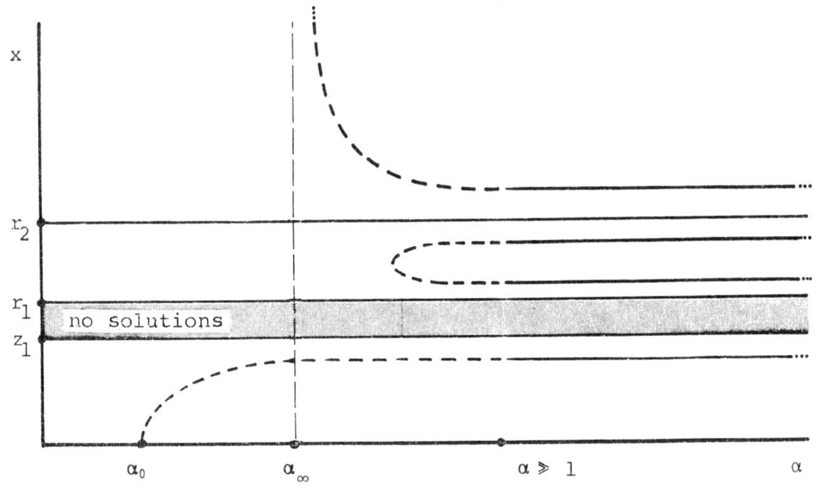

Figure 9. (case (1.10))

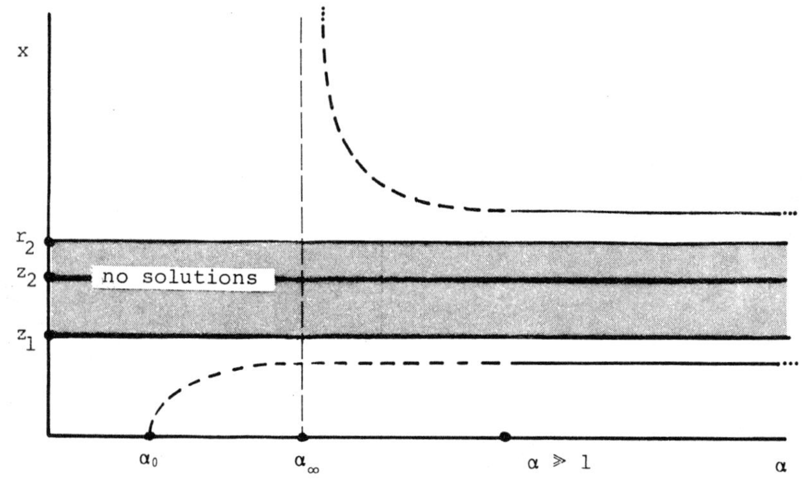

Figure 10. (case (1.11))

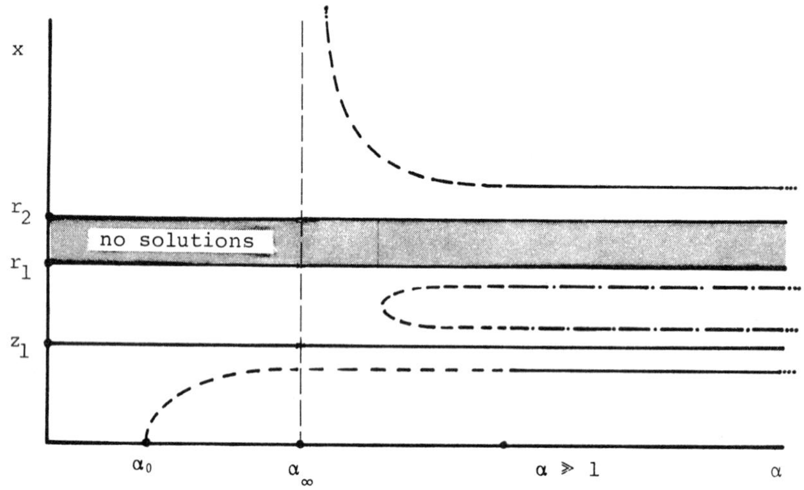

Figure 11. (case (1.12))

The broken lines in figures 9, 10 and 11 indicate how we expect that the 'fat' continua will continue as α becomes small. This is the content of section 2. There we will show bifurcation from zero at α_o and from infinity at α_∞ .

(3) We thus have that equation

$$\dot{x}(t) = - \alpha f(x(t-1))$$

has always <u>two</u> special periodic solutions for any $\alpha >> 1$ and in case

(1.10) we even can conclude <u>four</u> special periodic solutions for any $\alpha >> 1$.

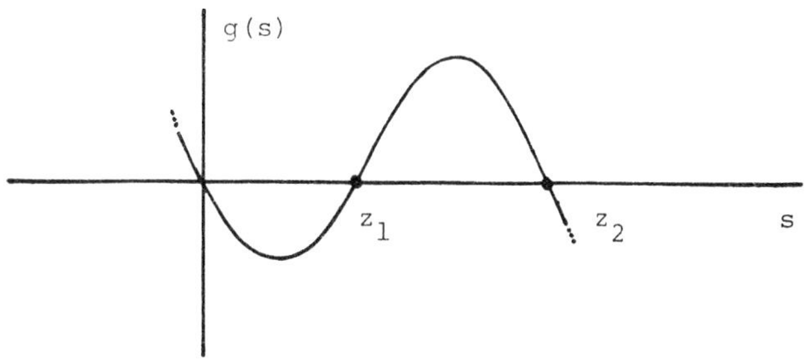

Figure 12.

(4) If we set $g = -f$ (cf. figure 12) we will have the same phase portraits as in figures 3, 5 and 7 <u>except that all arrows have to be reversed</u> . We thus have that equation

$$\dot{x}(t) = -\alpha g(x(t-1))$$

has

two special periodic solutions for any $\alpha >> 1$, provided that

$$\int_0^z g(s)ds > 0$$

and $-g$ satisfies the assumptions (1.2) .

These solutions are depicted in figure 11 by dotted lines.

(5) Figures 7 and 8 constitute explicit examples of typical <u>twists</u> in the sense of the Poincaré-Birkhoff fixed point theorem (cf. [24]) (see figures 13 and 14): for any $t \in R$ the phase flow Φ^t is an area preserving diffeomorphism with invariant curves S_1, S_2, S_3 and S_4 . S_1 and S_4 rotating counterclockwise and S_2 and S_3 rotating clockwise. Thus, the annuli A_1 (resp. A_2) constituted by S_1 and S_2 (resp. S_3 and S_4) are twists containing an even number of fixed points of Φ^t , one half elliptic and the other half hyperbolic. Indeed, as we saw, A_1 contains 4 hyperbolic and 4 elliptic rest

points and A_2 contains 8 hyperbolic and 8 elliptic rest points.

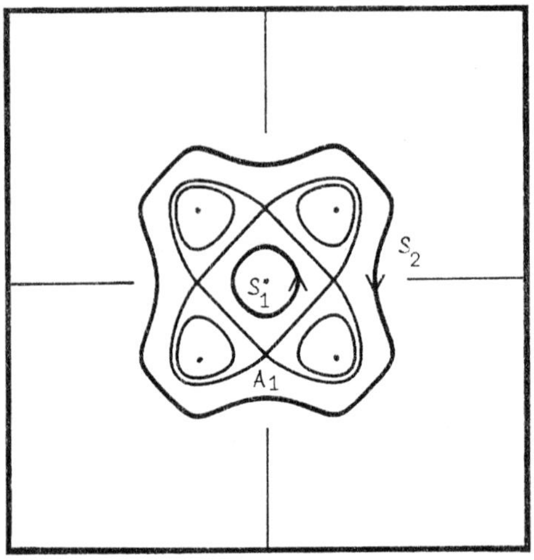

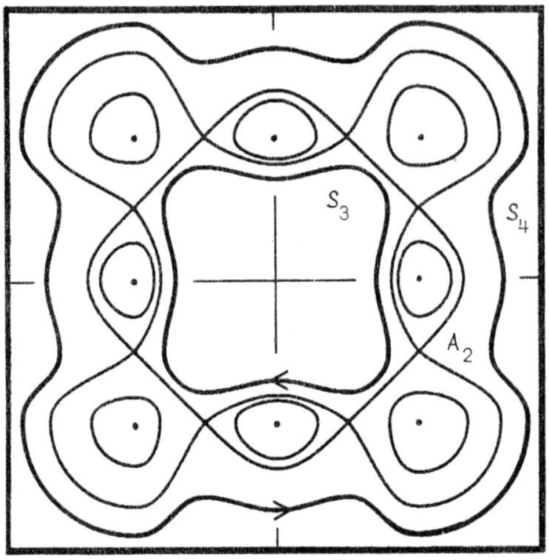

Figure 13. ($f=f_\varepsilon$ as in example 1.6 $\varepsilon= -5.0$) Figure 14.

We next want to study the phase portraits of the time-discrete dynamical system $T(f,h)$ given by (1.6). We want to study two classes of explicit examples for the generating nonlinearity f ; these will reveal structures which are typical for general functions f .

EXAMPLE 1.1. (C^1 model) Let $\varepsilon \in R$ and define

$$I_\varepsilon = \begin{cases} [\pi, 2\pi] & , \text{ if } \varepsilon \geq 0 \\ [0, \pi] & , \text{ if } \varepsilon \leq 0 \end{cases}$$

$$J_\varepsilon = \begin{cases} [0, \pi] & , \text{ if } \varepsilon \geq 0 \\ [\pi, 2\pi] & , \text{ if } \varepsilon \leq 0 \end{cases}$$

$$f_\varepsilon(s) = \begin{cases} \sin(s) & , s \in I_\varepsilon \\ \sin(s)+\varepsilon\sin^2(s) & , s \in J_\varepsilon \\ s - 2\pi & , s \geq 2\pi \end{cases}$$

and $f_\varepsilon(-s) = - f_\varepsilon(s)$. The perturbation term is chosen such that $f'_\varepsilon(z_i)$

(z_i a zero of f_ε) is independent of ε . This allows easier calculations. Observe that

$$\int_0^{z_2} f_\varepsilon(s)ds \begin{cases} > 0 \ , & \varepsilon > 0 \\ = 0 \ , & \varepsilon = 0 \\ < 0 \ , & \varepsilon < 0 \end{cases}$$

with $z_1 = \pi$, $z_2 = 2\pi$.

EXAMPLE 1.2. (PL model) Let ε, δ be given such that $0 \le \varepsilon \le 2$ and $0 \le |\delta| \le 2 - \varepsilon$. Define

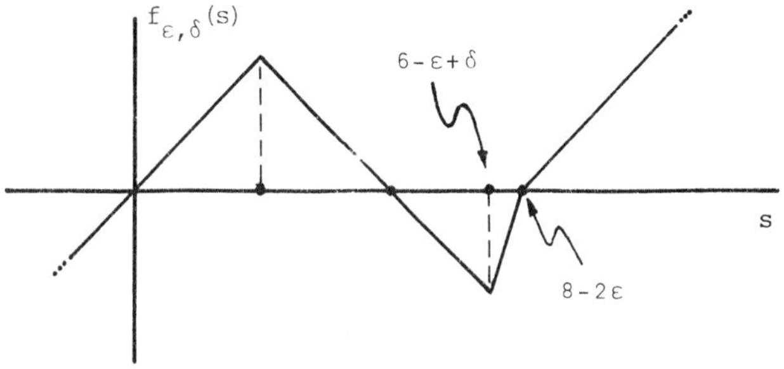

Figure 15.

$$f_{\varepsilon,\delta}(s) = \begin{cases} s & , \ 0 \le s \le 2 \\ -s + 4 & , \ 2 \le s \le 6 - \varepsilon + \delta \\ ms + b & , \ 6 - \varepsilon + \delta \le s \le 8 - 2\varepsilon \\ s - 8 + 2\varepsilon & , \ 8 - 2\varepsilon \le s \end{cases}$$

with $m = (2-\varepsilon+\delta)\ (2-\varepsilon-\delta)^{-1}$, $b = (2\varepsilon-8)m$, and $f_{\varepsilon,\delta}(-s) = - f_{\varepsilon,\delta}(s)$. Observe that

$$\int_0^{z_2} f_{\varepsilon,0}(s)ds \begin{cases} > 0 \ , & \varepsilon > 0 \\ = 0 \ , & \varepsilon = 0 \\ < 0 \ , & \varepsilon < 0 \end{cases}$$

with $z_1 = 4$ and $z_2 = 8 - 2\varepsilon$.

We first discuss some computer experiments for example (1.1). In figure 16 (i - v) we have $\varepsilon = 0$, which corresponds to case (1.11) . The discretization parameter h is

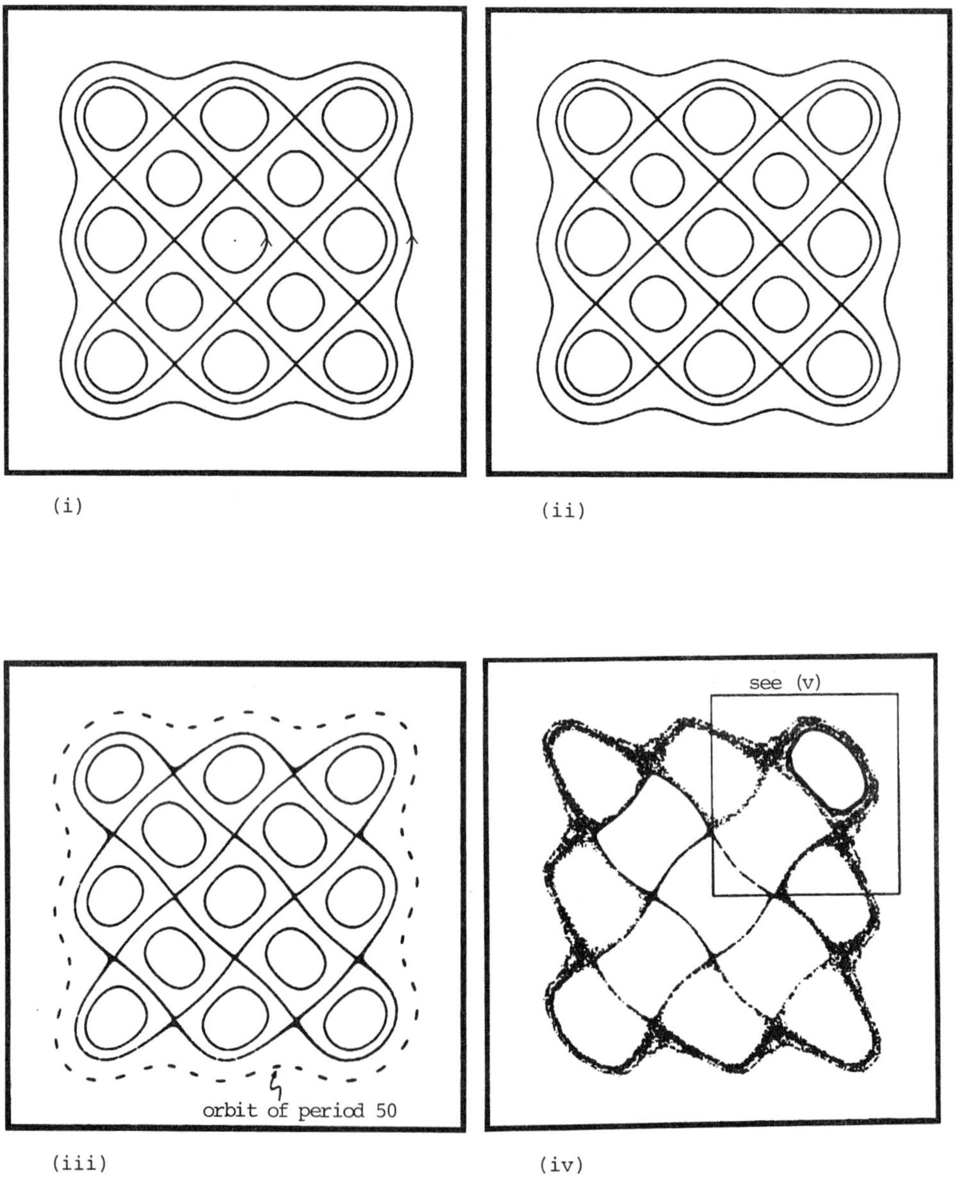

Figure 16. ($\varepsilon = 0$, $h = 0.05, 0.1, 0.5, 1.0$)

(i) : $h = 0.05$
(ii) : $h = 0.1$
(iii): $h = 0.5$
(iv) : $h = 1.0$, (v) is an enlargement of (iv)

The section of (x,y)-plane shown in (i-iv) is given by the window

$$\square = \{(x,y) : -15.0 \le x,y \le 15\} .$$

All figures are generated by plotting orbits of $T(f_\varepsilon,h)$ for particular initial values. As h increases, we observe the following facts:

-) The symmetries a,c and a_h , c_h (cf. (1.8)) differ by the perturbation term $hf(\cdot)$. While (i) still seems to reflect a and c , plots (ii - iv) obviously have lost these symmetries.

-) For h small it seems that the separatrices of the time-continuous system (1.1) are nicely approximated. In (iv) we have $3 \cdot 10^4$ iterations based on a single initial value close to the hyperbolic fixed point $(\pi,0)$ and we observe a "stochastic" distribution of points, reflecting that T corresponds to a surface of section of a non-integrable system.

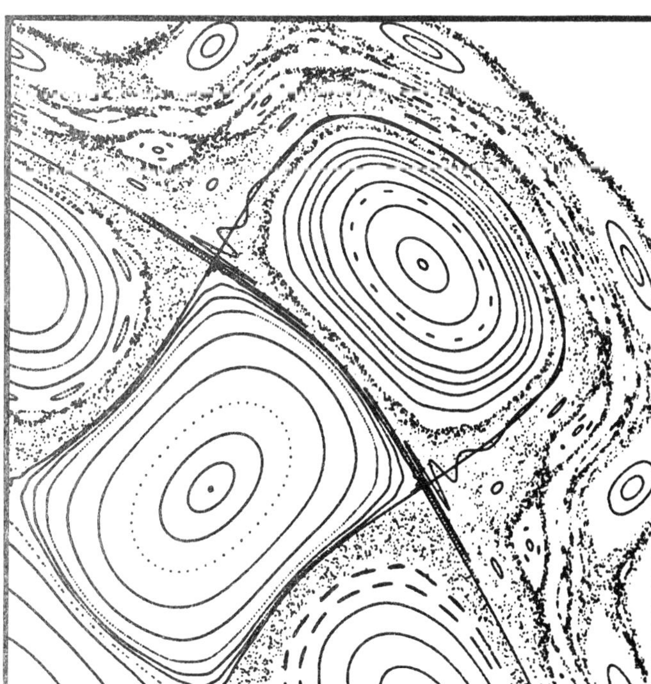

Figure 16. (v) (magnification of (iv))

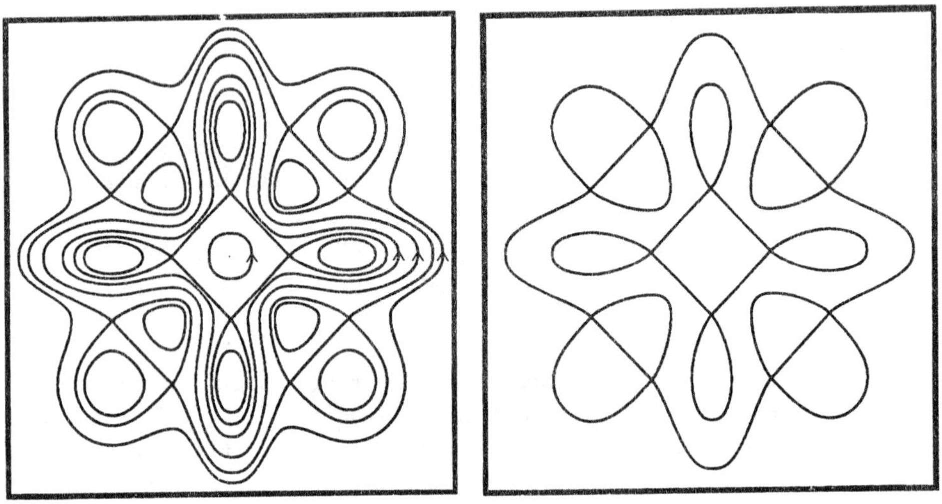

(i) (ii)

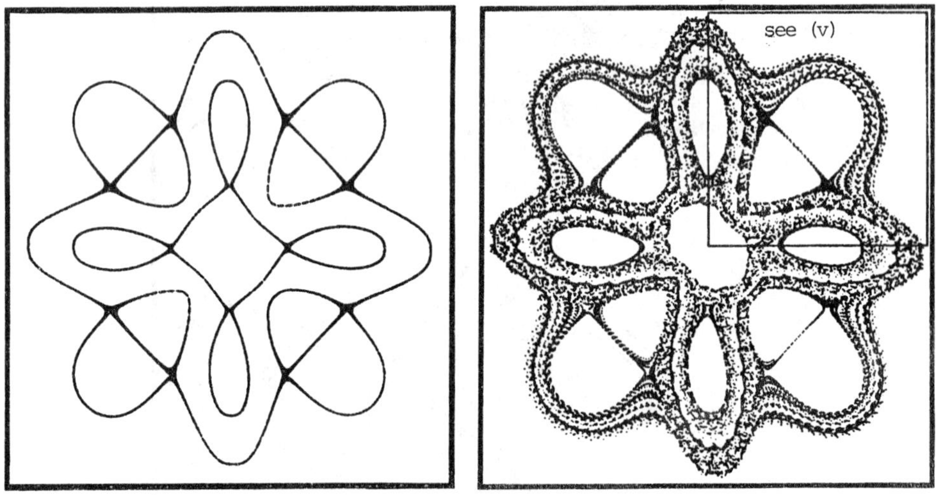

(iii) (iv)

Figure 17. ($\varepsilon = 5.0$, $h = 0.01, 0.01, 0.1, 0.2$)

This is due to the splitting of separatrices (existence of transversal homo- and heteroclinic points). In (v) we see the stable and unstable manifolds of the hyperbolic points $(2\pi,\pi)$ and $(\pi,2\pi)$ and observe that they intersect in transversal heteroclinic points and this yields the "chaotic dynamics" in a neighborhood of them. Experimentally there is evidence that this phenomenon is typical for any $h > 0$, however below visibility for h small in the scale chosen in (i-iii). We will in fact show the existence of homo- and heteroclinic points for all $h > 0$ in the case of special PL models.

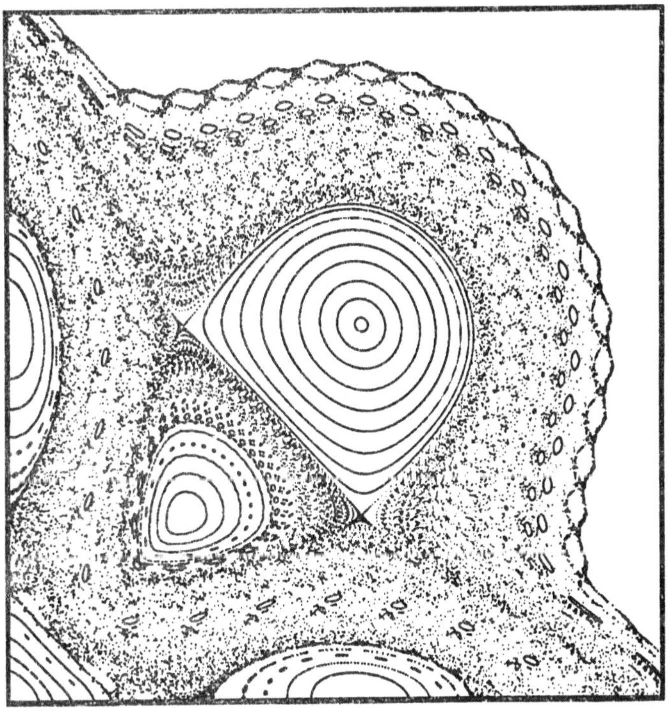

Figure 17. (v) (magnification of (iv))

Figure 17 is a series of plots for $\varepsilon = 5.0$, which corresponds to case (1.10). Again (iv) is obtained by $3 \cdot 10^4$ iterations of a single initial value close to $(\pi,0)$ and the windows are the same as before. Here the choices of h are:

(i), (ii) : $h = 0.01$
(iii) : $h = 0.1$
(iv) : $h = 0.2$

The effect of the ε-perturbation is here that the "stochastic" effects are amplified when compared with the plots for $\varepsilon = 0$.

(i)

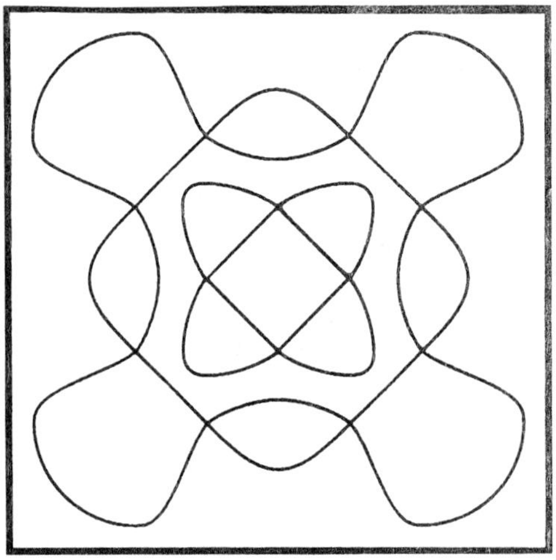

(ii)

Figure 18 is a series of plots for $\varepsilon = -5.0$, which correspond to case (1.12). Here the choices of h are:

(i), (ii) : $h = 0.01$
(iii) : $h = 0.1$
(iv) : $h = 0.2$
(vi) : $h = 0.3$

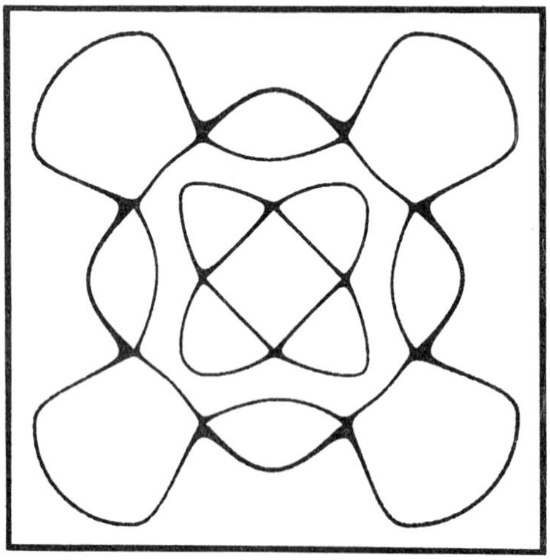

(iii)

Figure 18.

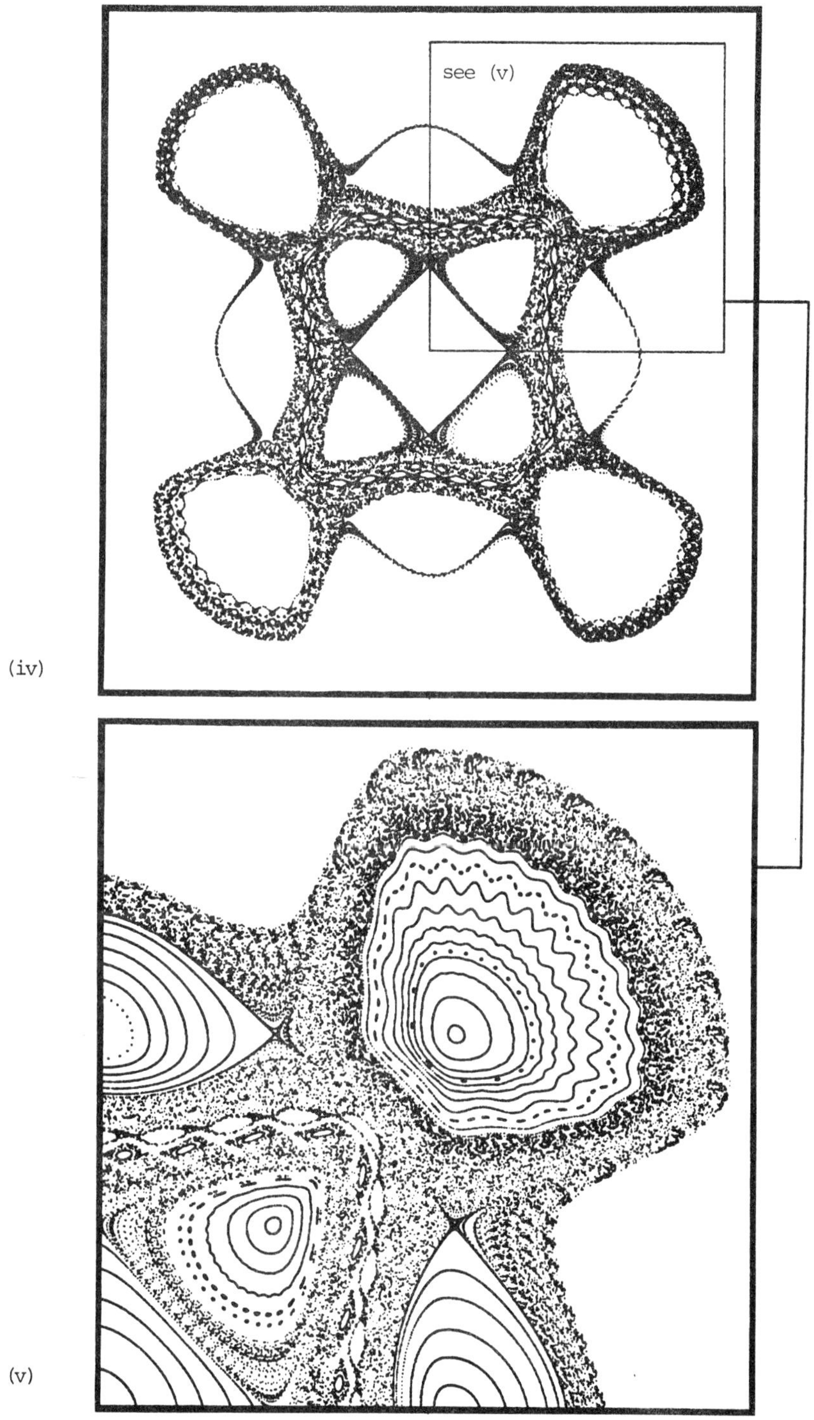

Figure 18.

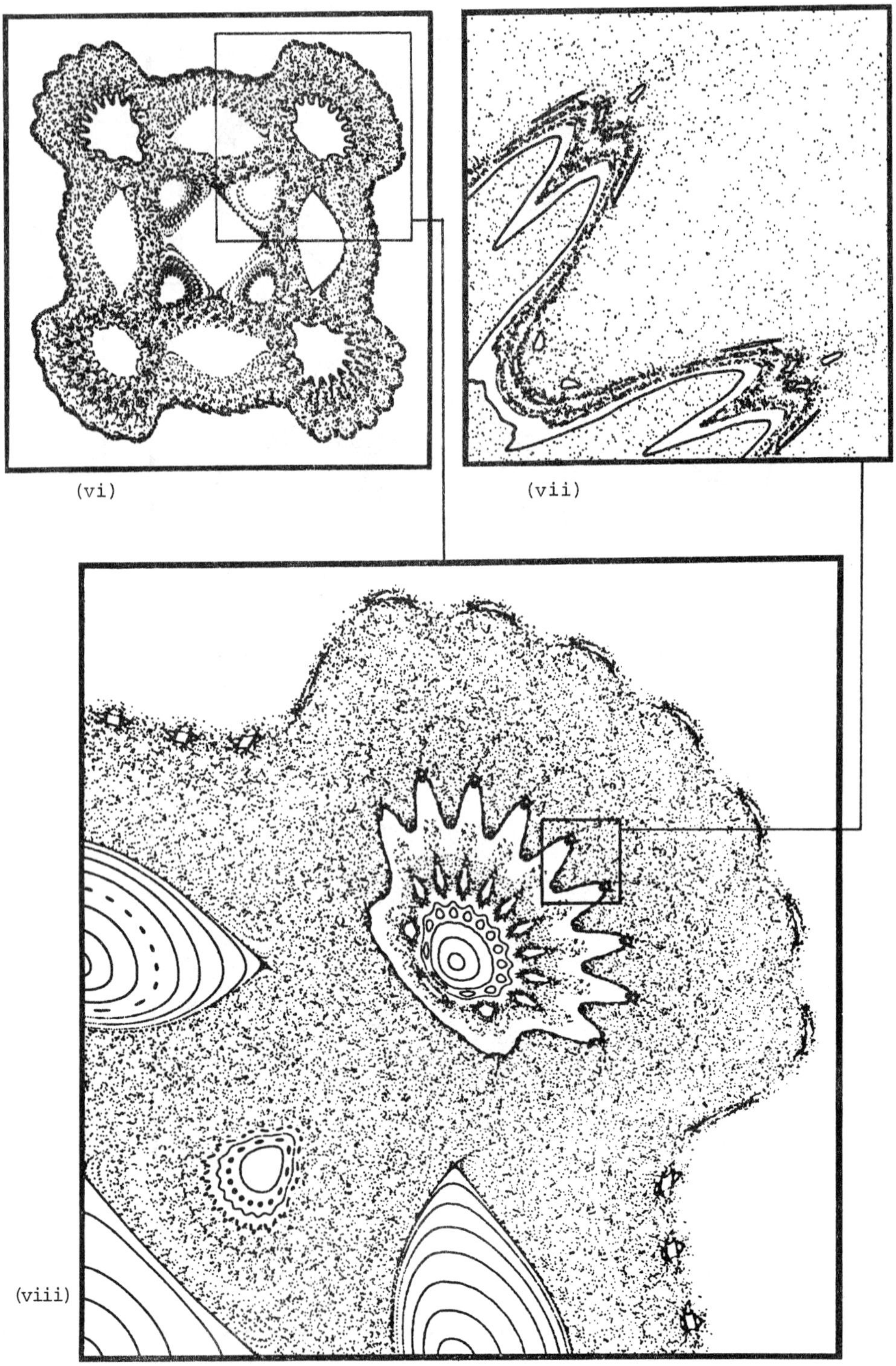

Figure 18. (vi),(vii),(viii)

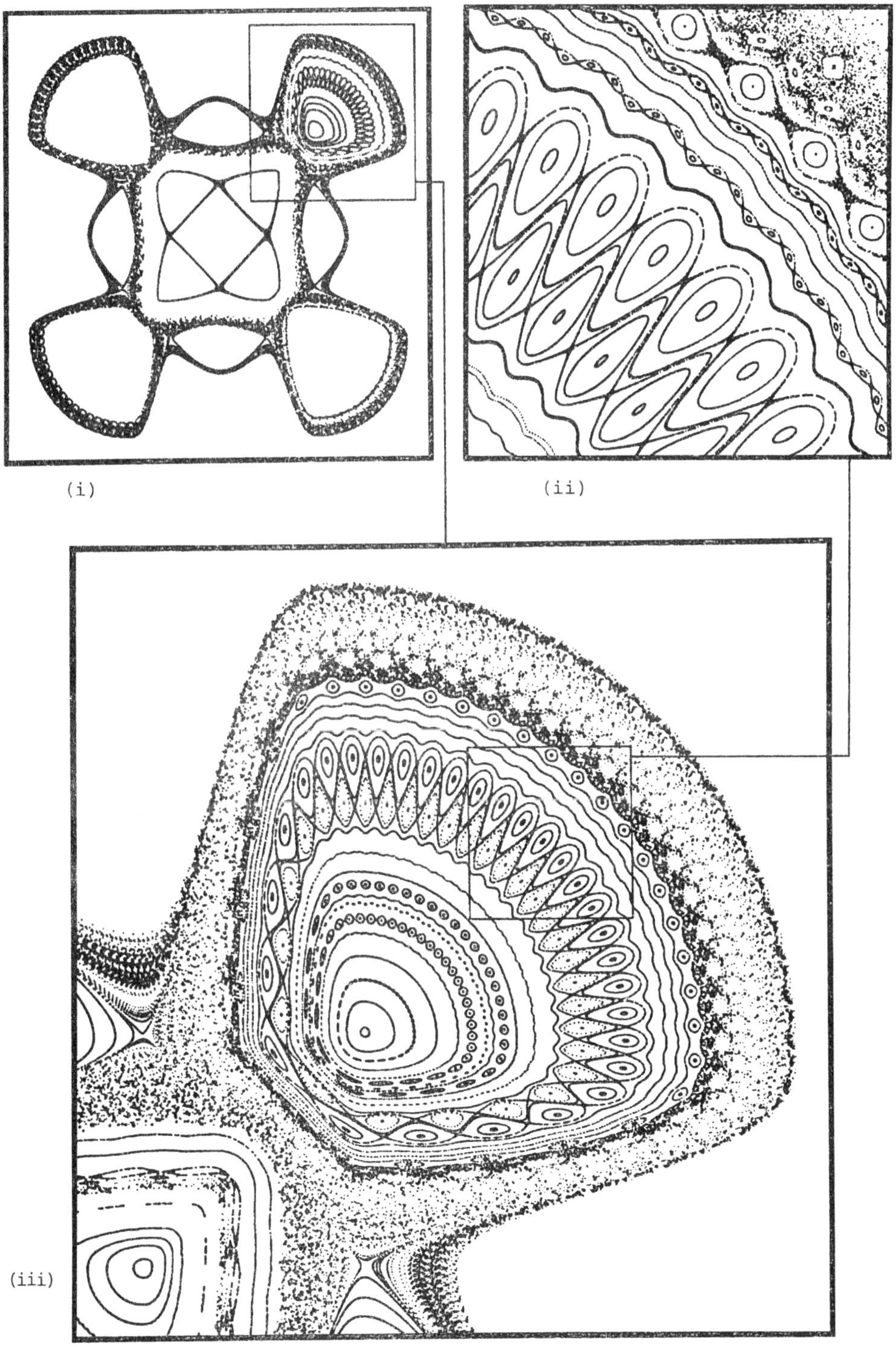

Figure 19. (ε = -10.0, h = 0.1)

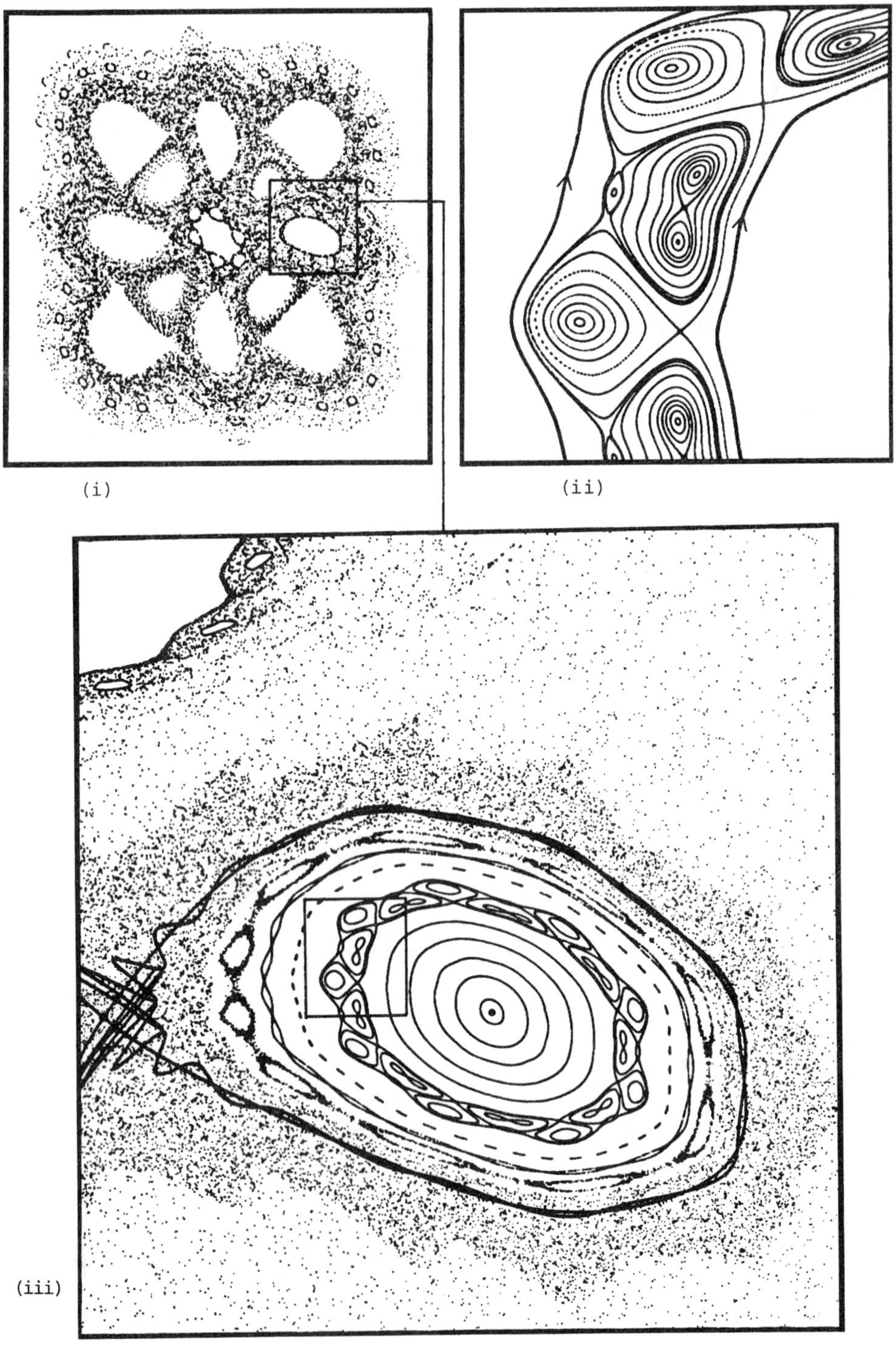

Figure 20. (ε = 2.0, h = 0.5)

Typically figures 18 (iv) and (vi) are again obtained by several ten thousands of iterations based on a single initial value close to the hyperbolic point $(\pi,0)$, while all magnifications in figures 16-18 are obtained by iteration from several particular initial values. Note that the invariant curves near elliptic points are due to one initial value. In figure 19 we show another model corresponding to case (1.12). Here $\varepsilon = -10.0$ and $h = 0.1$. The magnification (iii) reveals an interesting twist situation. Finally, in figure 20 we have $\varepsilon = 2.0$ and $h = 0.5$, i.e. it corresponds to case (1.10). In (ii) and (iii) we see magnifications of another interesting situation around the elliptic fixed point $(2\pi,0)$. In particular, (iii) suggests that we see the dynamics of a situation which has bifurcated from an "ordinary" twist situation, i.e. some elliptic periodic points have bifurcated into inverse hyperbolic periodic points yielding additional pairs of "new" elliptic periodic points of twice the "old" period.

The common feature of these experiments is that increasing ε or h results in an amplification of the erratic dynamics, i.e. more and more of the previously existing invariant curves around the elliptic points are destroyed. Both, with respect to the homoclinic regime and with respect to the elliptic regime (governed by the Kolmogorof-Arnold-Moser theory [22,24,46]) the experiments reveal such a variety of effects and phenomena that it appears hopeless to attempt to describe them rigorously. We therefore restrict attention in the following to the homoclinic and heteroclinic structure for specific models. Our first goal is to give an elementary discussion of the existence of <u>transversal</u> heteroclinic points, which is made possible by choosing the following model for a generating nonlinearity:

<u>EXAMPLE 1.3</u>. Let $0 < \delta << 1$ and $g_\delta : R \to R$ be a <u>smooth</u> function which is

$$g_\delta(s) = \begin{cases} s & , \ s \in [0,\ 2-\delta] \\ 4-s & , \ s \in [2+\delta\ ,\ 6-\delta] \\ s-8 & , \ s \in [6+\delta\ ,\ \infty)\ , \end{cases}$$

and $g_\delta(-s) = -g_\delta(s)$, and g_δ is sufficiently C^0-close to $f_{0,0}$, where $f_{0,0}$ is chosen according to example 1.2 (cf. figure 21) .

As a good representative for the hyperbolic structure we will study in the following the invariant manifolds of $P = (4,0)$ as families parametrized over ε , δ and μ . Since we will consider both g_δ and $f_{\varepsilon,\delta}$ as generating functions for T we will write $W^{s,u}(P;g_\delta)$ or $W^{s,u}(P;f_{\varepsilon,\delta})$ to indicate dependence on g_δ or $f_{\varepsilon,\delta}$. Note that for $\delta > 0$ $W^{s,u}(P;g_\delta)$ are immersed smooth 1-manifolds (cf. [38]) . As

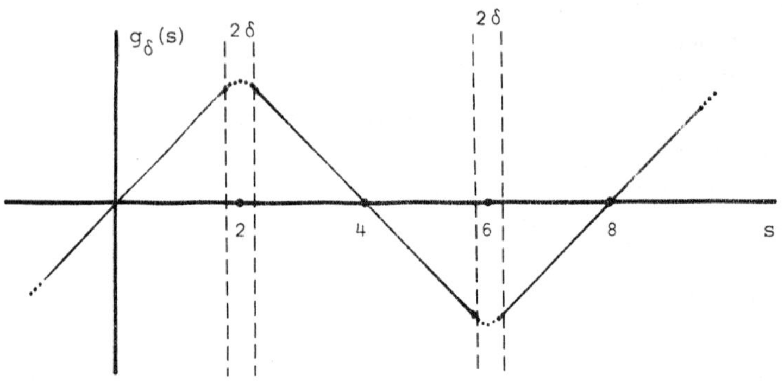

Figure 21.

$\delta \to 0$ it is reasonable to expect that one still has stable and unstable manifolds $W^{s,u}(P;g_o)$ which are approximated by $W^{s,u}(P;g_\delta)$. Here $W^{s,u}(P;g_o)$ naturally should be PL 1- manifolds. Indeed, for any generating function f which is PL the transformation T(f,h) is a PL homeomorphism, which is affine on pieces of R^2 . Moreover, if P is a fixed point of T and P is in the interior of $A \subset R^2$, where $T|_A$ is affine, and P is a hyperbolic fixed point, then

$$\begin{cases} W^u(P;f) = \bigcup_{n\geq 1} T^n\,(W^u_{loc}(P;f)) \\ W^s(P;f) = \bigcup_{n\geq 1} T^{-n}(W^s_{loc}(P;f)) \end{cases}$$

where

$$W^u_{loc}(P;f) = \{t\, e_u : t \in R\ ,\ |t| << 1\}$$

$$W^s_{loc}(P;f) = \{t\, e_s : t \in R\ ,\ |t| << 1\}$$

and e_u (resp. e_s) denote the unstable (resp. stable) eigenvector of T in P . Figure 23 ia a computer plot of the invariant manifolds for $T(g_\delta,h)$ for some $n \in N$ and <u>all</u> hyperbolic fixed points (g_δ as in example 1.3 ; $\delta = 0$, $h = 1.0$) . This choice corresponds to case (1.11), i.e. we expect a heteroclinic structure only. We will see however, that we also have homo-

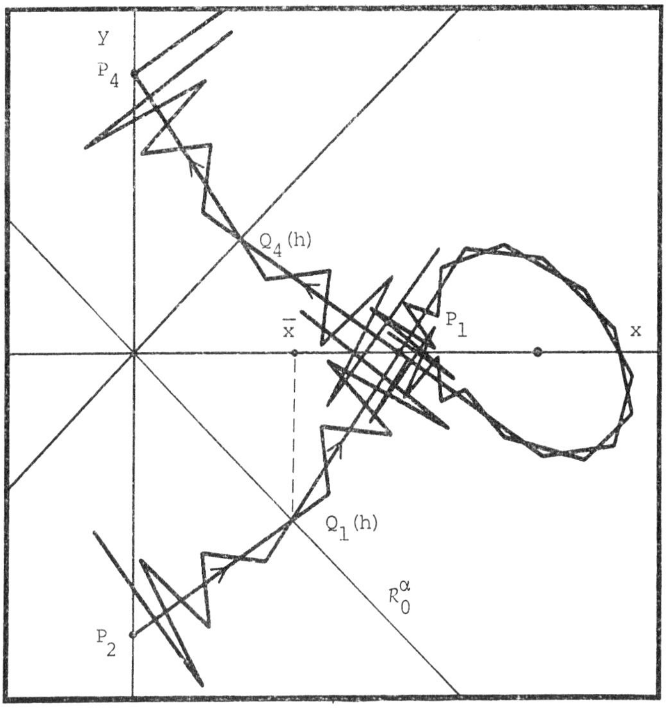

Figure 22. (homo- and heteroclinic structure, $\varepsilon=1.0$, $\delta=0$, $h=0.8$)

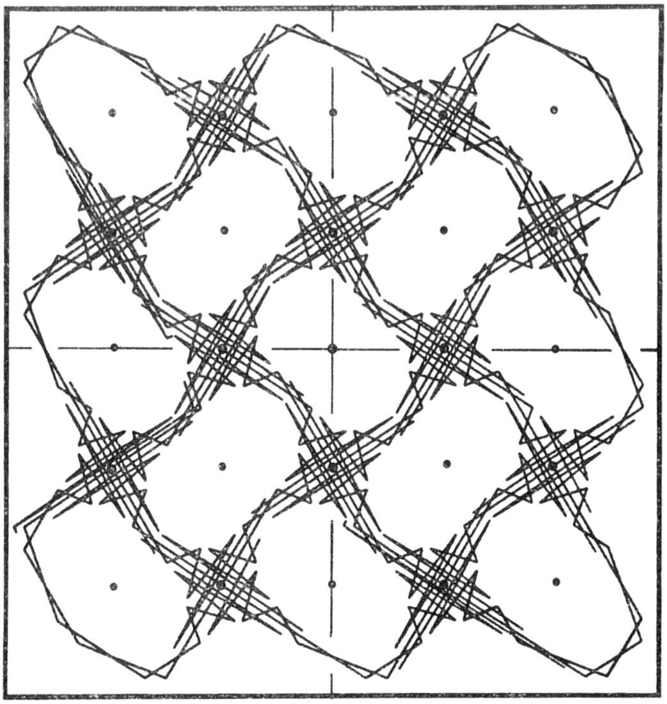

Figure 23. (all stable and unstable manifolds for $T(g_0,h)$, $h=1.0$)

clinic structure (cf. theorem 1.1). Figure 21 is a study for $T(f_{\varepsilon,\delta},h)$ ($f_{\varepsilon,\delta}$ as in example 1.2 ; $\varepsilon = 1.0$, $\delta = 0$, $h = 0.8$) . This corresponds to case (1.10), i.e. we expect from the discussion of the time-continuous system (1.1) the existence of a homoclinic structure and will see that this is true. Surprisingly the discussion of the heteroclinic structure is easy, while the existence of homoclinic points is much more delicate. In both cases we will make an essential use of the <u>isochrones</u> of T defined by the symmetries a_h , $b_h \equiv b$, $d_h \equiv d$ (cf. (1.8)) .

<u>DEFINITION 1.3</u>. (cf. [31,43]) Let $\alpha_h \in \{a_h , b_h , c_h , d_h\}$ be any of the involutions from (1.8) . Define the <u>isochrones</u> $R_n^{\alpha_h}$ for any $n \in Z$:

$$R_n^{\alpha_h} := \{(x,y) : R_n^{\alpha_h}(x,y) = (x,y)\} , R_n^{\alpha_h} = T^n\alpha_h .$$

Isochrones have played an essential role in [14,43] for the study of periodic points. Let P_n be the set of points of period n . Typically, the isochrones are diffeomorphic to the real line R , e.g. $R_o^{a_h}$ = x-axis . Their significance for the study of the homo-heteroclinic structure is that in view of lemma 1.2 it suffices to study $W^{s,u} \cap R_n^{\alpha_h}$ rather than $W^s \cap W^u$. To see that there is actually a reduction taking place we note that the isochrones are obtained from the basic isochrones $R_o^{\alpha_h}$ and $R_1^{\alpha_h}$:

<u>LEMMA 1.3</u>.

(i) $R_{2n}^{\alpha_h} = T^n(R_o^{\alpha_h})$, $R_{2n+1}^{\alpha_h} = T^n(R_1^{\alpha_h})$

(ii) $R_n^{\alpha_h} \cap R_m^{\alpha_h} \subset P_{n-m}$

<u>PROOF</u>. Suppose that $P \in R_o^{\alpha_h}$, so $\alpha_h(P) = P$ and $T^n\alpha_h(P) = T^n(P)$. Then we have $T^{2n}\alpha_hT^n(P) = T^{2n}\alpha_hT^n\alpha_h(P) = T^n(P)$. Conversely, if $P \in R_{2n}^{\alpha_h}$, one can check by similar arguments that $T^{-n}(P) \in R_o^{\alpha_h}$. Similarly, one can show that $T^n(R_1^{\alpha_h}) = R_{2n+1}^{\alpha_h}$. To prove (ii) , let $P \in R_n^{\alpha_h} \cap R_m^{\alpha_h}$ so $T^n\alpha_h(P) = P$ and $T^m\alpha_h(P) = P$, thus, $P = T^n\alpha_hT^m\alpha_h(P) = T^{n-m}(P)$.

Before we proceed to discuss the invariant manifolds of $P = (4,0)$ we introduce the following splitting for any hyperbolic fixed point $Q=(\bar{x},\bar{y})$.

If $\bar{x} > 0$ (resp. $\bar{x} < 0$) then let H_{in} be the half-space

$$H_{in} = \{x,y) : x \leq \bar{x} \quad (\text{resp. } x \geq \bar{x})\} .$$

If $\bar{x} = 0$ and $\bar{y} > 0$ (resp. $\bar{y} < 0$) then let

$$H_{in} = \{(x,y) : y \leq \bar{y} \ (\text{resp. } y \geq \bar{y})\} .$$

Denote by $H_{out} = \text{closure } (R^2 \backslash H_{in})$. Then define

$$\begin{cases} W^u_{in}(Q;f) = \bigcup_{n\geq 0} T^n(W^u_{1oc}(Q;f) \cap H_{in}) \\ W^s_{in}(Q;f) = \bigcup_{n\geq 0} T^{-n}(W^s_{1oc}(Q;f) \cap H_{in}) \end{cases}$$

etc.,

hence,

$$\begin{cases} W^s(Q;f) = W^s_{in}(Q;f) \cup W^s_{out}(Q;,f) \\ W^u(Q;f) = W^u_{in}(Q;f) \cup W^u_{out}(Q;,f) \end{cases}$$

For the discussion of the homo-heteroclinic structure of $P = (4,0)$ we will proceed by the following steps:

-) transversal heteroclinic intersections for the generating function g_δ;
-) transversal homoclinic intersections for the generating function g_δ;
-) transversal and degenerate intersections for the generating function $f_{\varepsilon,\delta}$.

We compute the stable (λ_s) and unstable (λ_u) eigenvalue of $T'(g_\delta,h)(P)$:

$$(1.19) \quad \begin{cases} \lambda_s = {}^1/_2(2+h^2-h\sqrt{h^2+4}) \\ \lambda_u = {}^1/_2(2+h^2+h\sqrt{h^2+4}) \end{cases}$$

Observe that $\lambda_s \cdot \lambda_u = 1$ and that for $h = 1$ one obtains

$$\lambda_{s,u} = {}^1/_2(3 \mp \sqrt{5})$$

which is the <u>golden mean</u>. The corresponding eigenvectors in the stable direction (e_s) and unstable direction (e_u) are

$$(1.20)\qquad \begin{cases} e_s = (1, h(1-\lambda_s)^{-1}) \\ e_u = (1, h(1-\lambda_u)^{-1}) \end{cases} .$$

Next let $\phi_h(x) = vx + w$ be the line through P with the slope of the stable eigenvector $e_s = e_s(h)$, i.e.

$$(1.21)\qquad \begin{cases} v = h(1-\lambda_s)^{-1} \\ w = -4h(1-\lambda_s)^{-1} \end{cases} .$$

We compute the point of intersection $Q = (\bar{x}(h), -\bar{x}(h))$ of the graph of ϕ_h with R_o^d to be

$$2 < \bar{x}(h) = 4h(1-\lambda_s+h)^{-1} < 4$$

for all $h > 0$. Observe that $\bar{x}(h) \to 2$ as $h \to 0$. Let $A \subset R^2$ be the set (see figure 24)

$$A = \{(x,y) : 2 \le x \le 4 \text{ and } hx - 4h - 2 \le y \le x - 4\}$$

$$\cup$$

$$\{(x,y) : 4 \le x \le 6 \text{ and } x - 4 \le y \le hx - 4h + 2\}$$

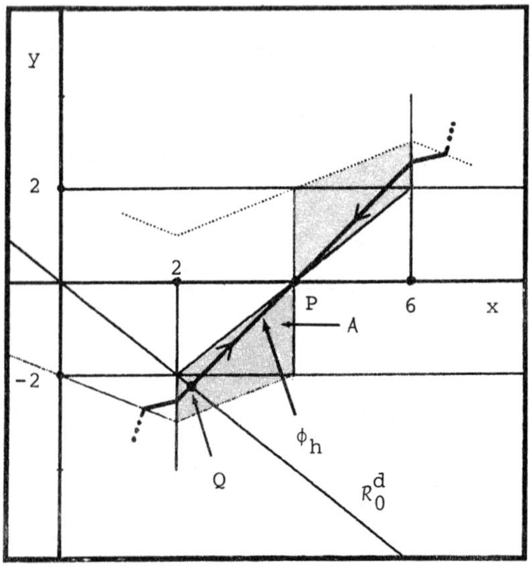

Figure 24.

One proves readily that for any $0 < h \leq 2$

$$T(g_o,h)_{|A} = L + \beta$$

where L is the linear map given by the matrix

$$\begin{pmatrix} 1+h^2 & -h \\ -h & 1 \end{pmatrix} \tag{1.22}$$

and $\beta^T = (-4h^2, 4h)$. One also checks directly that the graph of ϕ_h restricted to $x \in [2, 6]$ is entirely in A for $0 < h \leq 2$. Thus defining

$$L_\rho^s(h) = \{(x,y) : 2 + \rho \leq x \leq 4 - \rho \text{ and } y = \phi_h(x)\}$$

for $\rho \geq 0$, we have just proved that the line segment $L_o^s(h)$ has the property

$$L_o^s(h) \subset W_{in}^s(P;g_o) .$$

Using property (3) from lemma 1.2 we may conclude that

$$L_o^u(h) := d(L_o^s(h)) \subset W_{in}^u(d(P);g_o)$$

(d the involution from (1.8)) and therefore

$$Q = Q(h) = (\bar{x}(h), -\bar{x}(h)) \in R_o^d$$

is a transversal heteroclinic point for all $0 < h \leq 2$

$$Q \in W_{in}^s(P;g_o) \cap W_{in}^u(d(P);g_o) \cap R_o^d .$$

By arguments analogous to the above one proves that there exists a transversal heteroclinic point (c_o,b the involutions from (1.8))

$$S \in W_{in}^s(d(P);g_o) \cap W_{in}^u(c_o(P);g_o) \cap R_o^b$$

Applying lemma 1.2 again one obtains the transversal heteroclinic points (see figure 25)

$$b(Q) \in W^s_{in}(c(P);g_o) \cap W^u_{in}(b(P);g_o) \cap R^d_o$$

$$d(S) \in W^s_{in}(b(P);g_o) \cap W^u_{in}(P;g_o) \cap R^b_o .$$

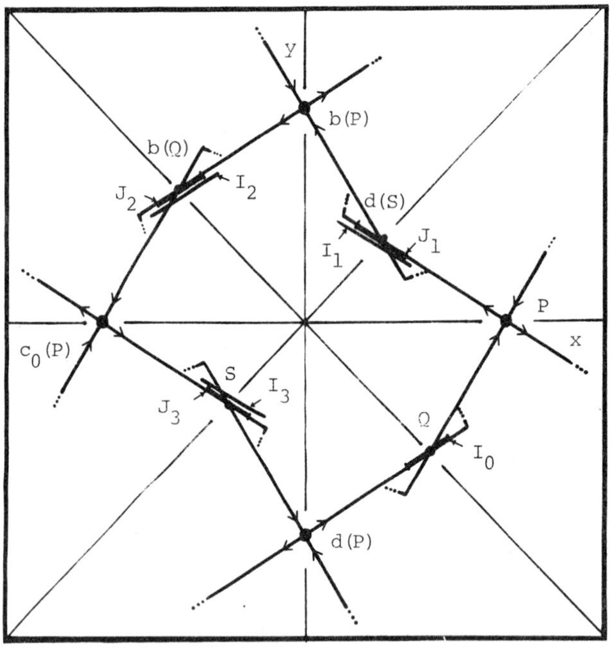

Figure 25.

Summarizing, we have the following heteroclinic structure

PROPOSITION 1.1. (C^∞ model)
For any h , $0 < h \le 2$, there exist

$$Q = Q(h) = (\bar{x}(h) , -\bar{x}(h))$$

$$S = S(h)$$

$$\delta_o > 0$$

such that for all δ , $0 < \delta < \delta_o$ one has the transversal heteroclinic intersections

$$\begin{cases} Q \in W^s_{in}(P;g_\delta) \cap W^u_{in}(d(P);g_\delta) \cap R^d_o \\ S \in W^s_{in}(d(P);g_\delta) \cap W^u_{in}(c_o(P);g_\delta) \cap R^b_o \\ b(Q) \in W^s_{in}(c_o(P);g_\delta) \cap W^u_{in}(b(P);g_\delta) \cap R^d_o \\ d(S) \in W^s_{in}(b(P);g_\delta) \cap W^u_{in}(P;g_\delta) \cap R^b_o \end{cases}$$

PROOF. The assertions for the smooth invariant manifolds $(g_\delta, \delta>0)$ follow from the assertions for the PL invariant manifolds (g_o) by the following elementary observation: Given h, $0 < h \le 2$, then there exist $\rho > 0$ and $\delta_o > 0$ such that

$$L^s_\rho(h) \subset W^s_{in}(P;g_\delta)$$

$$L^s_\rho(h) \cap R^d_o \neq \emptyset$$

for all $0 < \delta < \delta_o$; etc.

To discuss the homo-heteroclinic structure of $W^{s,u}_{out}(P,g_\delta)$, we need some notation:

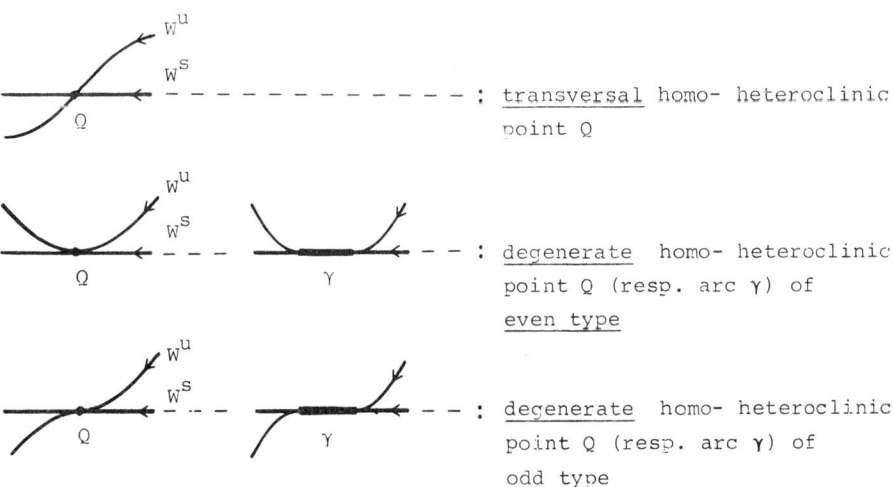

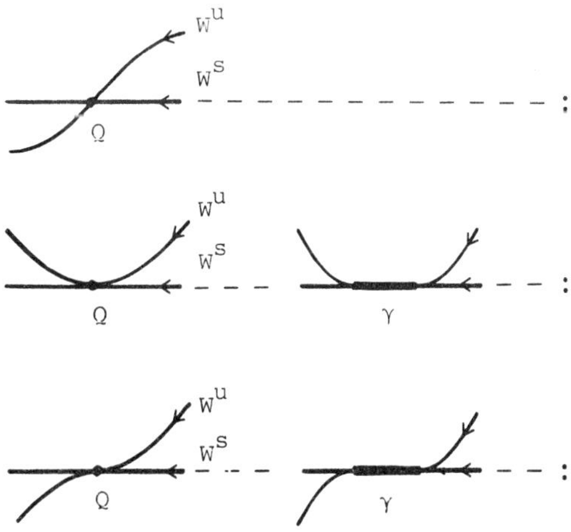

: transversal homo- heteroclinic point Q

: degenerate homo- heteroclinic point Q (resp. arc γ) of even type

: degenerate homo- heteroclinic point Q (resp. arc γ) of odd type

We will distinguish three types of homo-heteroclinic points (resp. arcs) according to the figure. There is one way that the degenerate even and odd type situations can coincide. This is when the common arc γ is the entire manifold $W^s = W^u$, e.g. in the integrable system (1.1). In their studies of the long-term stability for particle orbits L.J. Laslett and E.M. McMillan (see [22]) provided a whole class of area preserving diffeomorphisms with a totally degenerate heteroclinic structure (see figures 3o-37). One important aspect of such structures is that one may find stochastic dynamics that is confined to a finite region (bounded by the degenerate invariant manifolds), provided the diffeomorphism corresponds to a Poincaré section of a non-integrable system. S. Ushiki [41,42] has shown that for real analytic diffeomorphisms of R^2 which can be extended to automorphisms of C^2 the stable and unstable manifolds cannot have any regular arc in common.

Our next result discusses $W^{s,u}_{out}(P;g_\delta)$:

PROPOSITION 1.2. (C^∞ model)
For any h, $0 < h \le 2$, there exist $\bar{Q} = \bar{Q}(h)$, $\bar{S} = \bar{S}(h)$ and $\delta_o > 0$ such that for all δ, $0 < \delta < \delta_o$ one has the transversal heteroclinic intersections

$$\bar{Q} \in W^s_{out}(P,g_\delta) \cap W^u_{in}(U;g_\delta) \cap T_1$$

$$\bar{S} \in W^u_{out}(P,g_\delta) \cap W^s_{in}(a_o(U);g_\delta) \cap T_2$$

where $P = (4,0)$, $U = (8,4)$ and

$$T_1 = \{(x,y) : y = -x + 8\}, \quad T_2 = \{(x,y) : y = x - 8\}$$

(see figure 26) .

PROOF. We first discuss the case $\delta = 0$ and show the existence of $\bar{Q}$. We have already shown in the discussion of figure 24 that the line segment

$$\{(x,y) : y = \phi_h(x), \ 4 \le x \le 6\} \text{ is in } W^s_{out}(P;g_o) .$$

Now let $\psi_h(x) = vx + w$ be the line through U with

$$(1.23) \quad \begin{cases} v = h(\lambda_u - 1)^{-1} \\ w = [4(\lambda_u - 1) - 8h](\lambda_u - 1)^{-1} \end{cases}$$

i.e. ψ_h has the slope of the unstable eigenvector of $T'(g_\delta,h)(U)$. The graph of ϕ_h intersects the line T_1 in

$$\bar{Q} = \bar{Q}(h) = \left(\frac{8\lambda_s - 8 - 4h}{\lambda_s - 1 - h}, \frac{-4h}{\lambda_s - 1 - h}\right),$$

which is also the intersection of T_1 with the graph of ψ_h. Moreover the graphs of ϕ_h and ψ_h intersect in $\bar{Q}$ for all $0 < h \le 2$ transversally, and $\bar{Q} \to (6,2)$ as $h \to 0$, and the line segment between $\bar{Q}$ and U is in $W^u_{in}(U;g_o)$. Finally, the result follows for g_δ, δ sufficiently small, by the same argument as in proposition 1.1. The assertion for $W^u_{out}(P;g_\delta)$ and $W^s_{in}(a_o(u);g_\delta)$ is proved analogously.

So far our results discussing the heteroclinic structure for $T(g_\delta,h)$ were independent of h. This will not be true for the intersection properties of $W^s_{out}(U;g_\delta)$ and $W^u_{out}(a_o(U);g_\delta)$. However we have

PROPOSITION 1.3. (C^∞ model)
Let $h > h_2 = {}^1/_2\sqrt{2}$ and $h \neq h_1 = {}^1/_3\sqrt{12}$. Then there exist $\bar{R} = \bar{R}(h)$ and $\delta_o > 0$ such that one has the transversal heteroclinic intersection $\bar{R} \in W^s_{out}(U;g_\delta) \cap W^u_{out}(a_o(U);g_\delta) \cap R_o^{a_h}$, where $U = (8,4)$ (see figure 26).

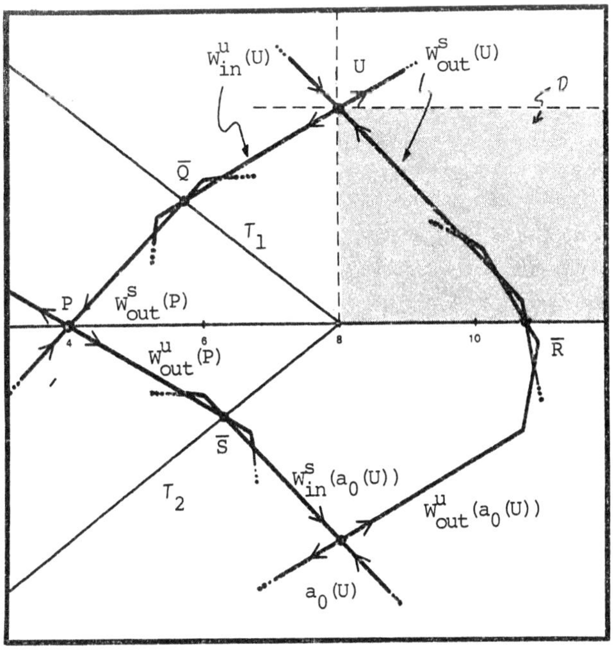

Figure 26.

Before we prove proposition 1.3 we introduce some useful notation: Let

$$(1.24)\qquad \begin{cases} \eta_k(x) = -hx + 8h + 6 - 4(k-1) \\ k = 1, 2, 3, 4 \end{cases}$$

be the lines bounding the strips

$$(1.25)\qquad \begin{cases} A_k = \{(x,y) : \eta_{k+1}(x) \le y \le \eta_k(x) \ , \ x \ge 6\} \\ k = 1, 2, 3 \end{cases}$$

Note that $T(g_o,h)(\{(x,y) : y = \eta_k(x) \ , \ x \ge 6\}) = \{(x,y) : y = 6 - 4(k-1), x \ge x_k\}$, where $x_1 = x_3 = 6 + 2h$, $x_2 = x_4 = 6 - 2h$. Also note that for all $h > 0$ $\eta_k(x) - \eta_{k+1}(x) = 4$ and that $U \in A_1$, $a_o(U) \in A_3$. Moreover we have that

$$T(g_o,h)|A_k \quad \text{is affine linear.}$$

PROOF. Let $\chi_h(x) = vx + w$ be the line through $U = (8,4)$ with the slope of the stable eigenvector of $T'(g_\delta,h)(U)$; i.e.

$$(1.26)\qquad \begin{cases} v = h(\lambda_s - 1)^{-1} \\ w = [4(\lambda_s-1) - 8h](\lambda_s-1)^{-1} \ . \end{cases}$$

Let (x^*,y^*) be the point of intersection of the graph of χ_h with the graph of η_2. Then one shows easily that $y^* = y^*(h) < 0$ for $h > h_2$ and the line segment between U and (x^*,y^*) is in A_1 and, thus, in $W^s_{out}(U;g_o)$ for all $h > 0$. Now let $h > h_2$ and let $\bar{R} = \bar{R}(h)$ be the point of intersection of the graph of χ_h with $R_o^{a_h}$ (i.e. the x-axis). One computes

$$\bar{R} = (8 - 4(\lambda_s - 1)h^{-1}, 0) \ .$$

Applying lemma 1.2 (property (3)) we conclude that

$$\bar{R} \in W^s_{out}(U;g_o) \cap W^u_{out}(a_o(U);g_o) \cap R_o^{a_h} \ .$$

For the PL model it remains to show that $\bar{R}$ is a transversal point of

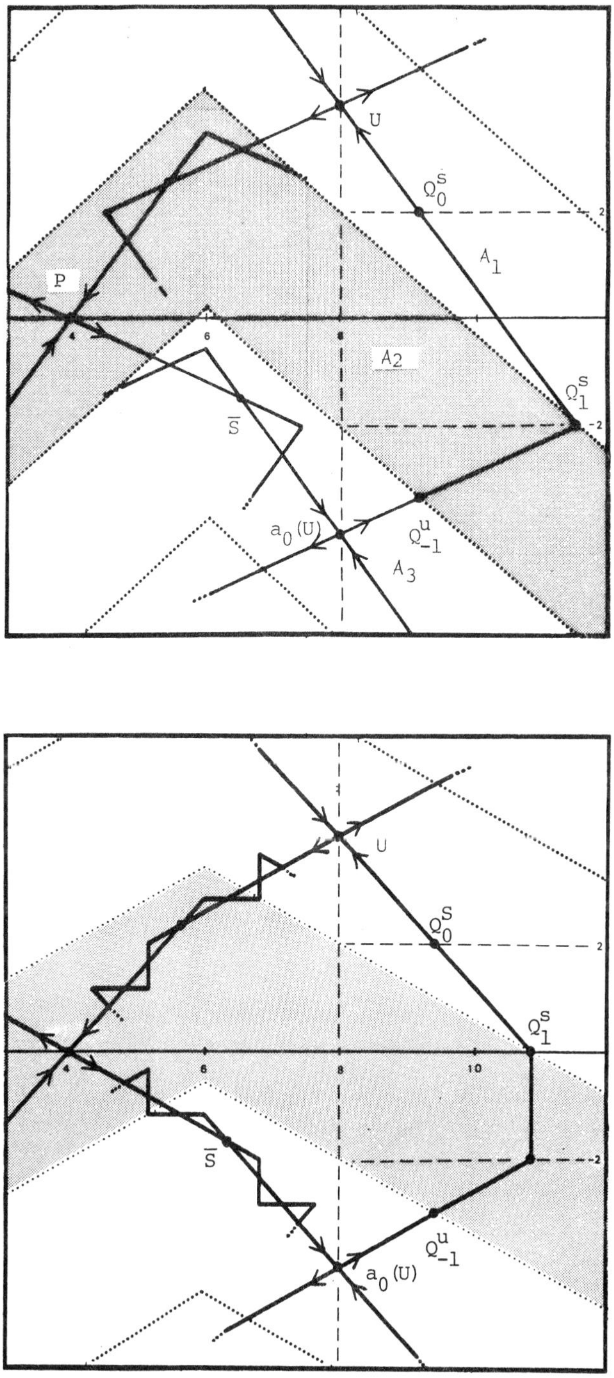

(a)
$h=h_1$

(b)
$h=h_2$

Figure 27. (singular heteroclinic structure)

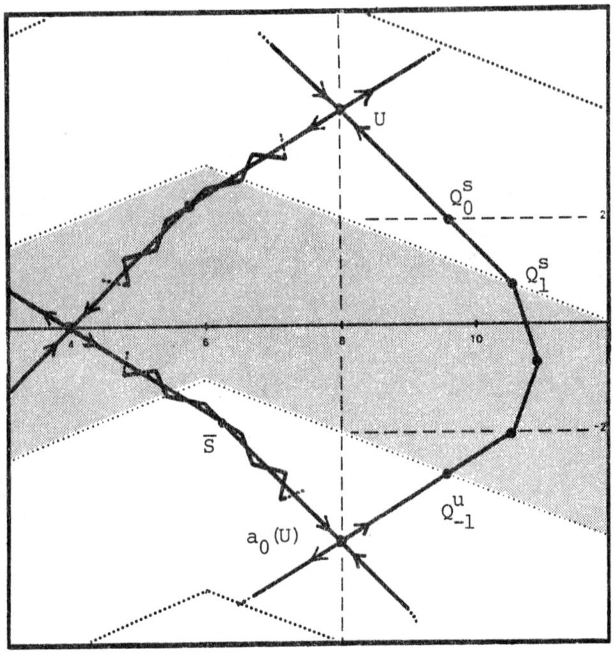

(c)
$h=h_3$

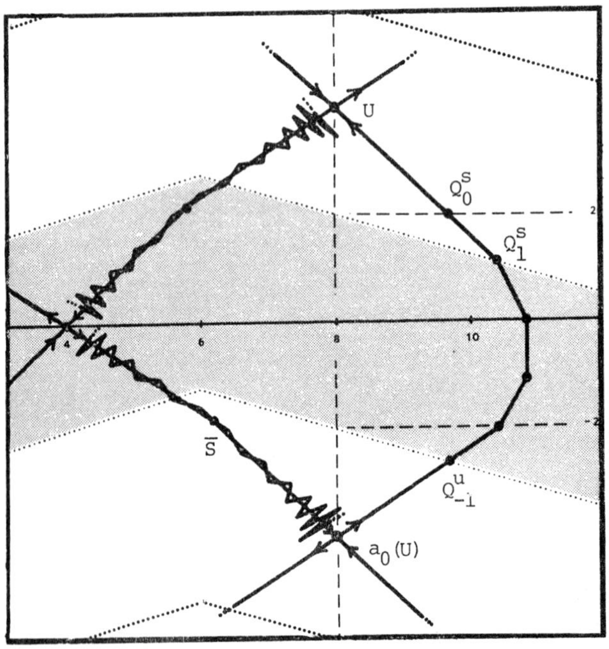

(d)
$h=h_4$

Figure 27. (singular heteroclinic structure)

intersection. Here we have to assume in addition to $h > h_2$ that $h \neq h_1 = {}^1/_3\sqrt{12}$. We study a small line segment $L_\varepsilon(h)$ on the graph of χ_h including $\bar{R}$ parametrized over $(\bar{R}_x - \varepsilon, \bar{R}_x + \varepsilon)$, where $\bar{R}_x$ denotes the x component of $\bar{R}$ and ε is small. Assume that $\bar{R}$ is not a transversal point of intersection. This would imply that for $\varepsilon << 1$ $a_h(L_\varepsilon(h)) \subset \mathrm{graph}(\chi_h)$. Consequently the slope of $L_\varepsilon(h)$ would be $-2h^{-1}$. However, the slope of χ_h is $h(\lambda_s - 1)^{-1}$. This implies a condition on h, which one computes to be $h = {}^1/_3\sqrt{12}$. Finally, we obtain the assertion for $\delta > 0$ by observing that the line segment

$$\{(x,y) : 8 \leq x \leq x^* - \rho, \; y = \chi_h(x)\}$$

is in $W^s_{out}(U;g_\delta)$ for $\rho > 0$ as small as we wish, provided δ is sufficiently small. So we may assume that $L_\varepsilon(h) \subset W^s_{out}(U;g_\delta)$ for $\varepsilon << 1$ and $a_h(L_\varepsilon(h)) \subset W^u_{out}(U;g_\delta)$, where $a_h(L_\varepsilon(h))$ is a small line segment due to the particular form of a_h.

The last proposition and its proof suggest that the one-parameter family $T(g_0,h)$ should encounter degenerations at $h = h_1$ and $h = h_2$. Indeed, we will show that there is a sequence $\{h_n\}_{n \in N}$ such that for $h = h_n$ we have the degenerate heteroclinic situation

$$W^s_{out}(U;g_0) = W^u_{out}(a_0(U);g_0),$$

which is familiar from integrable systems. Figure 27 shows computer experiments for

$$(1.27)\quad \begin{cases} (a)\ h = h_1 = {}^1/_3\sqrt{12} \approx 1.154701\ldots \\ (b)\ h = h_2 = {}^1/_2\sqrt{2} \approx 0.707106\ldots \\ (c)\ h = h_3 \approx 0.4975 \text{ (experimental value)} \\ (d)\ h = h_4 = {}^1/_2(3-\sqrt{5}) \approx 0.381966\ldots \end{cases}$$

<u>REMARK 1.3</u>. It is not hard to see that the structural properties of $W^s_{out}(U;g_0)$, $W^u_{out}(a_0(U);g_0)$ and $W^s_{out}(U_\varepsilon;f_{\varepsilon,\delta})$, $W^u_{out}(a_0(U_\varepsilon);f_{\varepsilon,\delta})$ are the same for all $h > 0$, $\varepsilon > 0$, $\delta = 0$ with $U_\varepsilon = (8-2\varepsilon,4)$. Thus, choosing smooth approximations to $f_{\varepsilon,\delta}$ for $\varepsilon > 0$ one will obtain transversal heteroclinic structures also for the case (1.10).

Before we proceed to discuss the degeneracies at $h = h_n$ we investigate the homoclinic structure of $W^s(P;g_\delta)$, $W^u(P;g_\delta)$. We will see that the <u>transversal</u> heteroclinic structure for these invariant manifolds implies also a transversal homoclinic structure and, thus, makes the Birkhoff-Smale theorem (see [4, 38]) applicable. We will obtain this structure essentially by an easy application of the λ-lemma due to J. Palis (see [25]) and proposition 1.1 . In the following theorem and its proof we use the notation of proposition 1.1 and let $X \in \{P, d(P), c_o(P), b(P)\}$.

THEOREM 1.1. (C^∞ model)

(1) For any h, $0 < h \le 2$, there exists $\delta_o > 0$, such that for any δ, $0 < \delta < \delta_o$, the invariant manifolds $W^s_{in}(X;g_\delta)$ and $W^u_{in}(X;g_\delta)$ have transversal homoclinic intersections.
(2) For any $h > h_2 = 1/2\sqrt{2}$ and $h \neq h_1 = 1/3\sqrt{12}$ there exists $\delta_o > 0$ such that for any δ, $0 < \delta < \delta_o$, the invariant manifolds $W^s_{out}(X;g_\delta)$ and $W^u_{out}(X;g_\delta)$ have transversal homoclinic intersections.

PROOF. We prove (1). The proof of (2) is analogous using proposition 1.2 and 1.3 . We choose δ_o according to proposition 1.1 and recall that $Q, b(Q)$, S, $d(S)$ are transversal heteroclinic points. An open 1-cell is by definition a differmorphic image of the interval $(0,1)$. We choose open 1-cells J_k, $k = 1,2,3,$ on (see figure 25)

J_1 on $W^u_{in}(P;g_\delta)$ containing $d(S)$

J_2 on $W^u_{in}(b(P);g_\delta)$ containing $b(Q)$

J_3 on $W^u_{in}(c_o(P);g_\delta)$ containing S .

Moreover, let I_0 be an open 1-cell on $W^u_{in}(X;g_\delta)$ containing Q, thus, intersecting $W^s_{in}(P;g_\delta)$ transversally in Q. We now apply the λ-lemma due to J. Palis (see [25], p. 22) three times. The statement of the λ-lemma is that

$$\bigcup_{n\ge 0} T^n(I_0)$$

contains 1-cells, which are arbitrarily C^1-close to J_1 and thus, intersect $W^s_{in}(b(P);g_\delta)$ transversally near $d(S)$. Let I_1 be one of these. Repeating the argument we thus obtain 1-cells I_2, which are arbitrarily C^1-close to J_2 and, finally, 1-cells I_3, which are arbitrarily C^1-close to J_3 and therefore must intersect $W^s_{in}(X,g_\delta)$ transversally near S. Note that $I_3 \subset W^u_{in}(X;g_\delta)$, which proves our assertion for $X = d(P)$. The homoclinic structures for the other choices of X are a simple consequence of lemma 1.2 .

We now discuss the degenerations of

$$W^s_{out}(U;g_o) \cap W^u_{out}(a_o(U);g_o)$$

for $h = h_n$ (see figure 27).

THEOREM AND CONJECTURE 1.2. (PL model)

(1) There exists a sequence $h_1 > h_2 > h_3 > \dots$ with $h_1 = {}^1/_3\sqrt{12}$, $h_2 = {}^1/_2\sqrt{2}$, $h_3 \approx 0.4975\dots$, $h_4 = {}^1/_2(3-\sqrt{5})$ such that for $h = h_n$, $n = 1,2,\dots$,

$$W^s_{out}(U;g_o) = W^u_{out}(a_o(U);g_o)\ , \quad \text{and} \quad h_n \to 0 \quad \text{as} \quad n \to \infty\ . \tag{1.28}$$

(2) Let $Q^s_o \in W^s_{out}(U;g_o)$ be the point

$$\left(8 - \frac{2(\lambda_s - 1)}{h}\ ,\ 2\right)$$

(i.e. the point of intersection of the graph of χ_h with the line $y \equiv 2$) .

Then

$$T^{-n}(Q^s_o) \in R^{a_h}_o\ , \tag{1.29}$$

provided $h = h_{2n}$, $n = 1,2,3\dots$.

(3) Let $m = -2h^{-1}$ and let L be the matrix

$$\begin{pmatrix} 1-h^2 & -h \\ h & 1 \end{pmatrix} \tag{1.30}$$

Furthermore, let $s_n = q_n/p_n$, where

$$L^{n-1}\begin{pmatrix} 1 \\ m \end{pmatrix} = \begin{pmatrix} p_n \\ q_n \end{pmatrix}\ .$$

Then

$$s_n = h(\lambda_s - 1)^{-1}$$

provided $h = h_{2n-1}$, $n = 1,2,3,\dots$.

(4) The degenerations of type (1.28) are the only possible ones for $W^s_{out}(U;g_0)$ and $W^u_{out}(a_0(U);g_0)$. Moreover, the degenerations are such that in the point K of first intersection of the PL manifold $W^s_{out}(U;g_0)$ with $R_o^{a_h}$ the manifold $W^s_{out}(U;g_0)$ has

- a breaking point, if $h = h_{2n}$,
- a line segment of slope m , if $h = h_{2n-1}$

(see figure 27 (a-d)) .

PROOF. We will prove (1), (2), (3) and (4) except for the existence of h_{2n-1} , $n = 3,4,\ldots$. Let $\mathcal{D} \subset R^2$ be the set

$$\mathcal{D} = \{(x,y) : x > 8 , 0 < y < 4\} .$$

Let $Q_1^s := T^{-1}(Q_o^s)$ (i.e. the point of intersection of the graphs of χ_h and η_2) . Note that the line segment between U and Q_1^s is on $W^s_{out}(U;g_0)$ and is contained in $\mathcal{D}$. Since

$$T^{-1}(x,y) = (x+hg_o(y) , y - hg_o(x+hg_o(y)))$$

it is easily seen that the y-components of $Q_k^s := T^{-k}(Q_o^s)$ are strictly decreasing for $Q_k^s \in \mathcal{D}$, as k increases, while the x-components are strictly increasing. Thus, given $h > 0$, there exists $k^* \geq 1$ such that

- either, $Q_{k^*}^s \in R_o^{a_h}$;
- or , the y-component of $Q_{k^*}^s$ is negative and the y-component of $Q_{k^*-1}^s$ is positive.

Moreover, it is obvious from the special form of T^{-1} that for each $Q_k^s \in \mathcal{D}$ the y-component of $Q_k^s = Q_k^s(h)$ is a decreasing and continuous function of h . To prove (1), (2) we now proceed as follows: We assume that we have $n \in N$ and $h > 0$ such that $T^{-n}(Q_o^s) \in R_o^{a_h}$ and show that this implies the degeneration (1.28). Then we show that there indeed exists a sequence h_{2n} , $n = 1,2,3, \ldots$, such that (1.29) holds. So let $h > 0$ be fixed and assume that $T^{-n}(Q_o^s) \in R_o^{a_h}$. Define $Q_{-1}^u \in W^u_{out}(a_0(U);g_0)$ to be the point of first intersection of $W^u_{out}(a_0(U);g_0)$ with η_3 . Note that the line segment between $a_0(U)$ and Q_{-1}^u is on $W^u_{out}(a_0(U);)$. Then set $Q_k^u = T^{k+1}(Q_{-1}^u)$, $k = 0,1,2,\ldots$. It will be useful to observe that

$$(1.31) \qquad a_h(Q_o^s) = Q_o^u .$$

In fact, one computes that

$$Q_o^u = \left(8 + \frac{2(\lambda_u - 1)}{h}, -2\right)$$

and applies a_h to see that (1.31) is true for all $h > 0$. Hence, using lemma 1.2 we have that

$$(1.32) \qquad a_h(T^n(Q_o^u)) = T^{-n}(Q_o^s) \in R_o^{a_h},$$

and, since $R_o^{a_h}$ is fixed under a_h, this implies that

$$(1.33) \qquad T^n(Q_o^u) = T^{-n}(Q_o^s).$$

Now let L_{gen}^s be the line segment (on $W_{out}^s(U;g_o)$) between Q_o^s and Q_1^s and let L_{gen}^u be the line segment (on $W_{out}^u(a_o(U);g_o)$) between Q_{-1}^u and Q_o^u. Note that

$$W_{out}^s(U;g_o) = \bigcup_{k\geq 0} T^k(L_{gen}^s) \cup \bigcup_{k\geq 0} T^{-k}(L_{gen}^s)$$

$$W_{out}^u(a_o(U);g_o) = \bigcup_{k\geq 0} T^{-k}(L_{gen}^u) \cup \bigcup_{k\geq 0} T^k(L_{gen}^u).$$

Thus, it suffices to show that

$$(1.34) \qquad T^n(L_{gen}^u) = T^{-n}(L_{gen}^s).$$

Note, however, that $L_{gen}^u \subset A_3$ and $T(L_{gen}^u), \ldots, T^n(L_{gen}^u) \subset A_2$ (see (1.25)) and therefore $T^n(L_{gen}^u)$ is itself a line segment. Similarily it follows that $T^{-n}(L_{gen}^s)$ is a line segment. Finally, (1.34) then follows from (1.33). This proves (1.28) under the assumption (1.29). We now prove the existence of a sequence h_{2n}, $n = 1,2,3, \ldots$, satisfying (1.29). The values h_2 and h_4 can be explicitly computed as zeros of polynomials in h (see figure 27(b) and (d)) using the explicit formula for L_{gen}^s. For $n > 2$ one could in principle carry out an explicit calculation, however, the degree of the corresponding polynomials increases with n and therefore a general argument seems to be more adequate.

Assume that h_{2n} is found already such that (1.29) is satisfied and Q_n^s has a x-component greater than 8 (this is true for $n = 1,2$). We argue that there exists h_{2n+1} with $h_{2n+1} < h_{2n}$ satisfying (1.29). Recalling the particular form of T^{-1} on $\mathcal{D}$ we may conclude that the y-component of $Q_{n+1}^s(h)$, which is negative, will increase as $h < h_{2n}$ decreases and that the x-component of $Q_{n+1}^s(h)$ is bounded from below by 8 and decreasing as $h < h_{2n}$ decreases. Since T^{-1} is close to Id for $0 < h << 1$ we can argue that the y-component of $Q_{n+1}^s(h)$ is positive for h sufficiently small. Thus, by continuity we find $h = h_{2n+1} < h_{2n}$ such that $Q_{n+1}^s(h_{2n+1}) \in R_o^{a_h}$. The assertion that $h_{2n} \to 0$ as $n \to 0$ follows because the y-component of T^{-1} is decreasing on $\mathcal{D}$. This proves (1) for n even. To prove (4) assume that $h = h_*$ is such that (1.28) is satisfied. Let K be the point of first intersection of $W_{out}^s(U;g_o)$ with $R_o^{a_h}$. Then either K is a breaking point of the PL manifold $W_{out}^s(U;g_o)$ or there is a small line segment $L_K(h_*)$ on $W_{out}^s(U;g_o)$ including K, which is invariant under a_h (see lemma 1.2). Hence, in the latter case the slope of $L_K(h_*)$ is $m = -2h_*^{-1}$. Now in A_2 $T(g_o,h)$ is given by $L + \beta$, where L is the linear map (1.30). Let $n - 2$ be the unique integer such that

$$T^{n-1}(L_K(h_*)) \subset A_1$$

$$T^{n-2}(L_K(h_*)) \subset A_2 \quad .$$

Since $T^{n-1}(L_K(h_*)) \subset W_{out}^s(U;g_o)$ if then follows that $s_n = h_*(\lambda_s - 1)^{-1}$, i.e. the slope of the stable eigenvector of $T'(g_o,h_*)(U)$, where $s_n = q_n/p_n$ and

$$L^{n-1}\begin{pmatrix}1\\m\end{pmatrix} = \begin{pmatrix}p_n\\q_n\end{pmatrix} \quad .$$

For $n = 1$ this leads to a polynomial equation of degree 2 in h which provides h_1. For $n = 2$ one obtains already a polynomial equation of degree 6

$$8h^6 - 48h^4 + 76h^2 - 16 = 0$$

which is equivalent to

$$u^3 - 12u^2 + 38u - 16 = 0 \; ; \qquad u = 2h^2 \; .$$

The cubic polynomial has precisely one real root u_*, and $u_* > 0$. Then

$$h_3 = {}^1/_2 \sqrt{2u_*} \approx 0.4975\ldots$$

If K is a breaking point of $W^s_{out}(U;g_o)$ then again one argues with the affine linear structure in A_2 to conclude that there exists an n such that $T^{n-1}(g_o,h)(K)$ is a point on the graph of η_2 bounding A_2, and this puts us in the situation (1.29).

REMARK 1.4. (1) Using a PL-version of the λ-lemma of J. Palis a consequence of theorem 1.2 is that $W^s_{out}(P;g_o)$ and $W^u_{out}(P,g_o)$ $(P=(4,0))$ have a transversal homoclinic structure for any $h > 0$. If $h \neq h_n$, $n = 1, 2, \ldots,$ one can argue similarily to the proof of theorem 1.1, using the fact that $W^s_{out}(U;g_o)$ and $W^u_{out}(a_o(U);g_o)$ have a transversal intersection on $R^{a_h}_o$. If, however, $h = h_n$ one can use the fact $W^s_{out}(U;g_o) = W^u_{out}(a_o(U);g_o)$ to show that a 1-cell on $W^u_{out}(P;g_o)$ including $\bar{S}$ (see prop. 1.2) under iteration of T will be mapped arbitrarily close to a 1-cell on $W^u_{out}(U;g_o)$, which intersects $W^s_{out}(P;g_o)$ transversally.

(2) The degenerations of type (1.28) in theorem 1.2 also may occur in C^∞ models. For example consider $h = h_2$ and $W^s_{out}(U;g_o) = W^u_{out}(a_o(U);g_o)$. One easily constructs a C^∞ function g_δ in the class of examples 1.3 for which $W^s_{out}(U;g_\delta) = W^u_{out}(a_o(U);g_\delta)$. This is done in the following way: Let

$$\eta_k(x,\delta) = -hx + 8h + 6 - 4(k-1) + \delta, \quad \delta > 0$$

and

$$B_k(\delta) = \{(x,y) : x \geq 6, \ \eta_k(x,-\delta) \leq y \leq \eta_k(x,\delta)\}$$

We want to construct g_δ in such a way that

$$L_o := W^s_{out}(U;g_o) \cap R^2 \setminus (B_2(\delta) \cup B_3(\delta) \cup T(B_3(\delta)))$$
$$=$$
$$W^s_{out}(U;g_\delta) \cap R^2 \setminus (B_2(\delta) \cup B_3(\delta)) \cup T(B_3(\delta))).$$

for $\delta > 0$ sufficiently small. So let C be a smooth 1-cell which can be parametrized by a function $\gamma(x)$, where γ is invertible, and which can be attached to L_o to fill in the gap at $T(B_3(\delta))$ smoothly. Let E be the line segment

$$W^s_{out}(U;g_o) \cap T(B_2(\delta)).$$

Then g_δ is defined by the functional equation

$$a_h(C) = E \quad \text{(see figure 28)} .$$

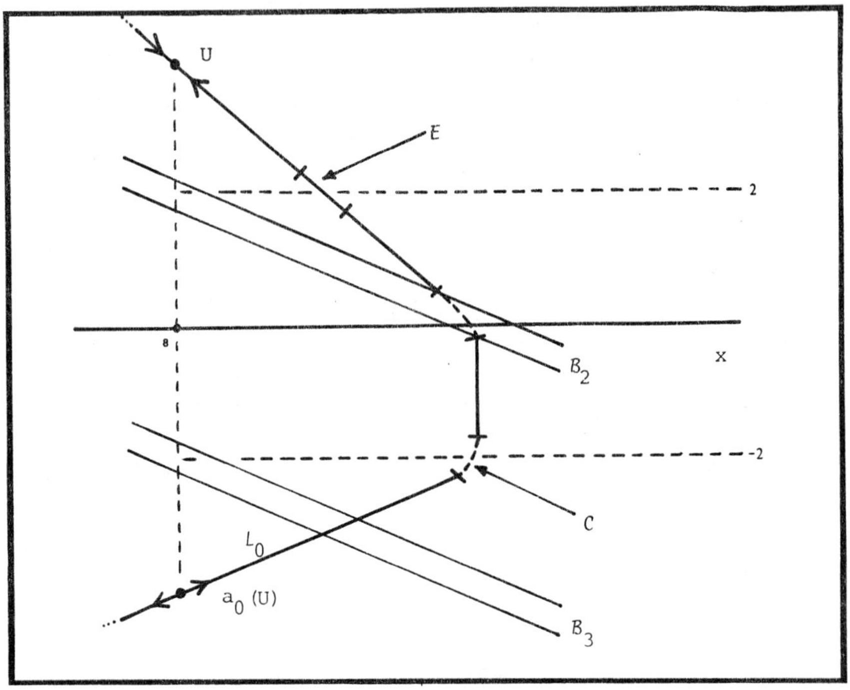

Figure 28. ($h = h_2$)

Thus, for example, for $\gamma(x) \in [-2-\delta,-2+\delta]$ one obtains

$$g_\delta(\gamma(x)) = - 1\{(\gamma(x)+4)(\lambda s-1)h^{-2} + (x-8)h^{-1}\} .$$

We now proceed to discuss the homoclinic structure attached to the fixed point $P = (4,0)$ for the generating nonlinearity $f_{\varepsilon,\delta}$, where ε,δ are such that

$$\int_0^{8-2\varepsilon} f_{\varepsilon,\delta}(s)ds > 0 .$$

This corresponds to case (1.10), i.e. the phase flow of (1.1) has a homoclinic orbit. Note that (see cases (1.10)-(1.12)) for any $t > 0$ the diffeomorphism Φ^t has a degeneration of type (1.28). Thus $T(g_0,h)$ reflects this property for any $h = h_n$. We will see in the following that $T(f_{\varepsilon,\delta},h)$ also reflects for particular choices of h the homoclinic orbit of Φ^t attached to P (case (1.10)). In general, however, this homoclinic orbit of Φ^t breaks into a transversal homoclinic structure of the approximating T .

It will be useful to specify the affine linear structure for $T(f_{\varepsilon,\delta},h)$: Let

$$(1.35)\qquad \begin{cases} \eta_k^\varepsilon(x) := -hx + (8-2\varepsilon)h + 6 - 4(k-1) \\ \zeta_k(x) := hx - 4h + 6 - 4(k-1) \end{cases}$$

$$k = 2,3$$

and

$$(1.36)\qquad \begin{cases} A_l^\varepsilon = \{(x,y) : 2 \le x \le 6-\varepsilon \text{ and } \zeta_3(x) \le y \le \zeta_2(x)\} \\ A_r^\varepsilon = \{(x,y) : 6-\varepsilon \le x \text{ and } \eta_3^\varepsilon(x) \le y \le \eta_2^\varepsilon(x)\} . \end{cases}$$

Note that $T(f_{\varepsilon,\delta},h)|_{A_l^\varepsilon}$ and $T(f_{\varepsilon,\delta},h)|_{A_r^\varepsilon}$ are affine linear maps for $\delta = 0$.

PROPOSITION 1.4. (PL model)
Let $f_{\varepsilon,\delta}$ be as in example 1.2 with $\varepsilon = 1$ and $\delta = 0$. Let $h = \sqrt{2}$. Set $Q = (5, h(1-\lambda_s)^{-1})$. Then K , the intersection of the line segment from $a_h(Q)$ to $T(a_h(Q))$ with $R_o^{a_h}$ is a transversal homoclinic point for $T = T(f_{\varepsilon,\delta},h)$, i.e.

$$K \in W^s_{out}(P;f_{\varepsilon,\delta}) \cap W^u_{out}(P;f_{\varepsilon,\delta}) .$$

PROOF. (see figure 29) Our argument is similar to that in the proof of proposition 1.1. Firstly, the line segment between P and Q is in A_l^ε , $\varepsilon = 1$, and is therefore on $W^s_{out}(P;f_{\varepsilon,\delta})$. Now one computes

$$\lambda_s = 2 - \sqrt{3} ,\ Q = (5, {}^1/_2(\sqrt{6} + \sqrt{2})) .$$

Applying lemma 1.2 one obtains that the line segment between P and $a_h(Q)$ is on $W^u_{out}(P;f_{\varepsilon,\delta})$ and one computes that

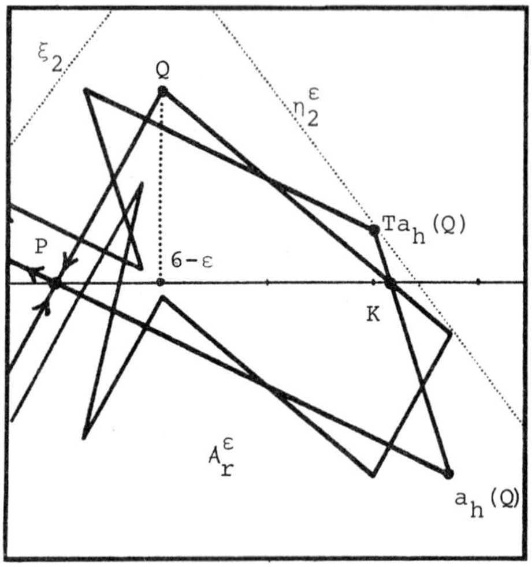

Figure 29. (transversal homoclinic structure)

$$a_h(Q) = (6 + \sqrt{3}\,,\ -\,{}^1/_2(\sqrt{6} + \sqrt{2}\,))$$

and

$$T(a_h(Q)) = (7\ ,\ {}^1/_2(\sqrt{6} - \sqrt{2}\,))\ .$$

Moreover the line segment L from $a_h(Q)$ to $T(a_h(Q))$ is on $W^u_{out}(P;f_{\varepsilon,\delta})$ and intersects $R_o^{a_h}$ in K . The slope of L is $-\sqrt{6}\,(\sqrt{3}-1)^{-1}$ and this is different from $-2h^{-1}$. Therefore $a_h(L) \subset W^s_{out}(P;f_{\varepsilon,\delta})$ intersects L transversally in K .

We see that the idea to obtain a <u>homoclinic point</u> is to show that (due to lemma 1.2) $W^u_{out}(P;f_{\varepsilon,\delta})$ intersects $R_o^{a_h}$ in a point K . Then <u>transversality</u> follows, provided K is a point of a line segment L on $W^u_{out}(P;f_{\varepsilon,\delta})$ such that L intersects $a_h(L)$ transversally. Thus, conditions for a totally degenerate homoclinic structure are intersections such that

(1.37)
- (i) K is a breaking point of $W^u_{out}(P;f_{\varepsilon,\delta})$, or
- (ii) K is a point of a line segment L on $W^u_{out}(P;f_{\varepsilon,\delta})$ with slope $-2h^{-1}$.

For the following result see figures 30-38.

THEOREM AND CONJECTURE 1.3. (PL model)

(1) There exists a sequence $\mu_1 > \mu_2 > \mu_3 > \ldots, \mu_n \to 0$ as $n \to 0$, with

$\mu_1 = {}^1/_2\sqrt{7 + \sqrt{17}} \approx 1.6675\ldots$, $\mu_2 = {}^1/_2\sqrt{4 + 2\sqrt{2}} \approx 1.3065\ldots$,

$\mu_3 \approx 1.0521\ldots$, $\mu_4 \approx 0.8736\ldots$, $\mu_5 \approx 0.7439\ldots$,

$\mu_6 \approx 0.6465\ldots$, $\mu_7 \approx 0.5707\ldots$, $\mu_8 \approx 0.5106\ldots$,

such that for $h = \mu_n$, $n = 1, 2, 3, \ldots$

$$W^s_{out}(P;f_{\varepsilon,\delta}) = W^u_{out}(P;f_{\varepsilon,\delta}) \,, \quad \delta = 0 \,, \tag{1.38}$$

for any $\varepsilon \in (0,2)$ such that

$$\mu_n \le \rho(\varepsilon) \,,$$

where $\rho(\varepsilon) = (4\varepsilon - \varepsilon^2)(4 - 2\varepsilon)^{-1}$.

(2) Let $Q^s_o \in W^s_{out}(P;f_{\varepsilon,\delta})$ be the point

$$(6 - \varepsilon \,,\ (2 - \varepsilon)h(1 - \lambda_s)^{-1}) \,,$$

(i.e. the point of intersection of the graph of ϕ_h (see (1.21)) with the line $x \equiv 6 - \varepsilon$) . Let $\delta = 0$ and $\varepsilon \in (0,2)$. Then

$$T^{-n}(Q^s_o) \in R_o^{a_h} \tag{1.39}$$

provided $h = \mu_{2n}$, $n = 1, 2, 3, \ldots$. and $\mu_{2n} \le \rho\ (\varepsilon)$.

(3) Let $m = -2h^{-1}$ and let L be the matrix

$$\begin{pmatrix} 1-h^2 & -h \\ h & 1 \end{pmatrix} \tag{1.40}$$

Furthermore, let $s_n = q_n/p_n$, where

$$L^n \begin{pmatrix} 1 \\ m \end{pmatrix} = \begin{pmatrix} p_n \\ q_n \end{pmatrix} .$$

Then

$$s_n = h(1 - \lambda_s)^{-1}$$

provided $h = \mu_{2n-1}$, $n = 1, 2, 3 \ldots$, for any $\varepsilon \in (0,2)$ such that $\mu_{2n-1} \leq \rho(\varepsilon)$.

(4) The degenerations of type (1.38) at $h = \mu_n$ are the only possible ones for $W^s_{out}(P;f_{\varepsilon,\delta})$, $W^u_{out}(P;f_{\varepsilon,\delta})$, $\delta = 0$, provided $\mu_n \leq \rho(\varepsilon)$. Moreover the degenerations are of type (1.37 i) for $h = \mu_{2n}$ and of type (1.37 ii) for $h = \mu_{2n-1}$.

<u>PROOF</u>. The proof is essentially analogous to the proof of theorem 1.2, except that it is more difficult here to show that $W^s_{out}(P;f_{\varepsilon,\delta})$, $\delta = 0$, intersects $R_o^{a_h}$ for all $h > 0$. Also the critical parameter values $\mu_1, \mu_2, \mu_3, \ldots$ are zeroes of polynomials of higher degree (see (1.39) and (1.40)). The value for μ_2, for example, is obtained as follows: Let $h = \mu_2$ be such that $T^{-1}(Q_o^s)$ is on $R_o^{a_h}$, i.e. has y-component equal to zero. One computes

$$Q_o^s(h) = (6-\varepsilon, z) = (6-\varepsilon, \tfrac{2-\varepsilon}{2}(h+\sqrt{h^2+4})), \quad z = z(h),$$

thus, $0 \leq z \leq 2$ for all $h > 0$ such that $h \leq \rho(\varepsilon)$. Choosing ε sufficiently close to 2, h is allowed to be as large as we wish. Then $T^{-1}(f_{\varepsilon,\delta},h)(Q_o^s)$, $\delta = 0$, has the y-component

$$-h^2 z + z + h(2-\varepsilon),$$

and, thus, property (1.39) is independent of ε for $n = 1$, as long as $h \leq \rho(\varepsilon)$. This yields $\mu_2 = {}^1/_2\sqrt{4 + 2\sqrt{2}}$. In the proof of theorem 1.2 we made an essential use of the fact that the y-components of $Q_k^s \in \mathcal{D}$ are decreasing as k increases. Let

$$\bar{\mathcal{D}} = \{(x,y) : x > 8 - 2\varepsilon, \ 0 < y < 4\}, \quad Q_k^s := T^{-k}(Q_o^s).$$

(a)

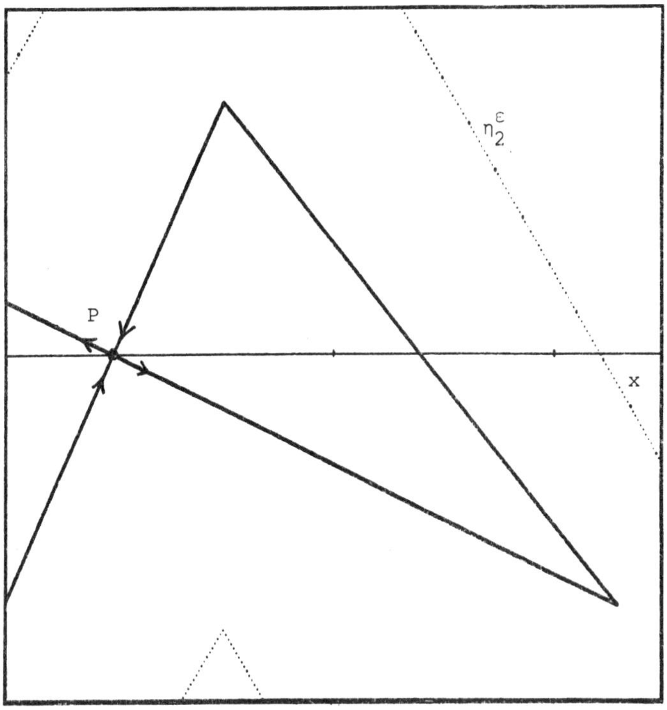

(b)

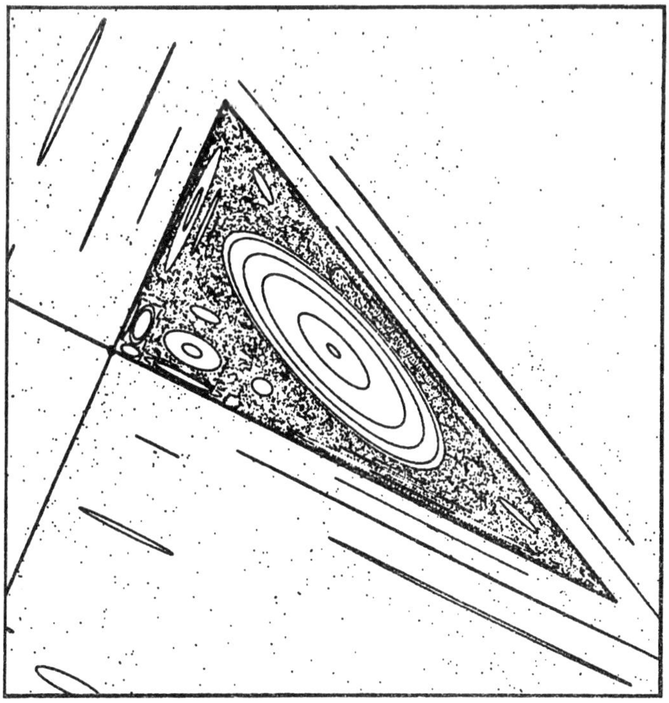

Figure 30. (singular separatrix, $h = \mu_1$)

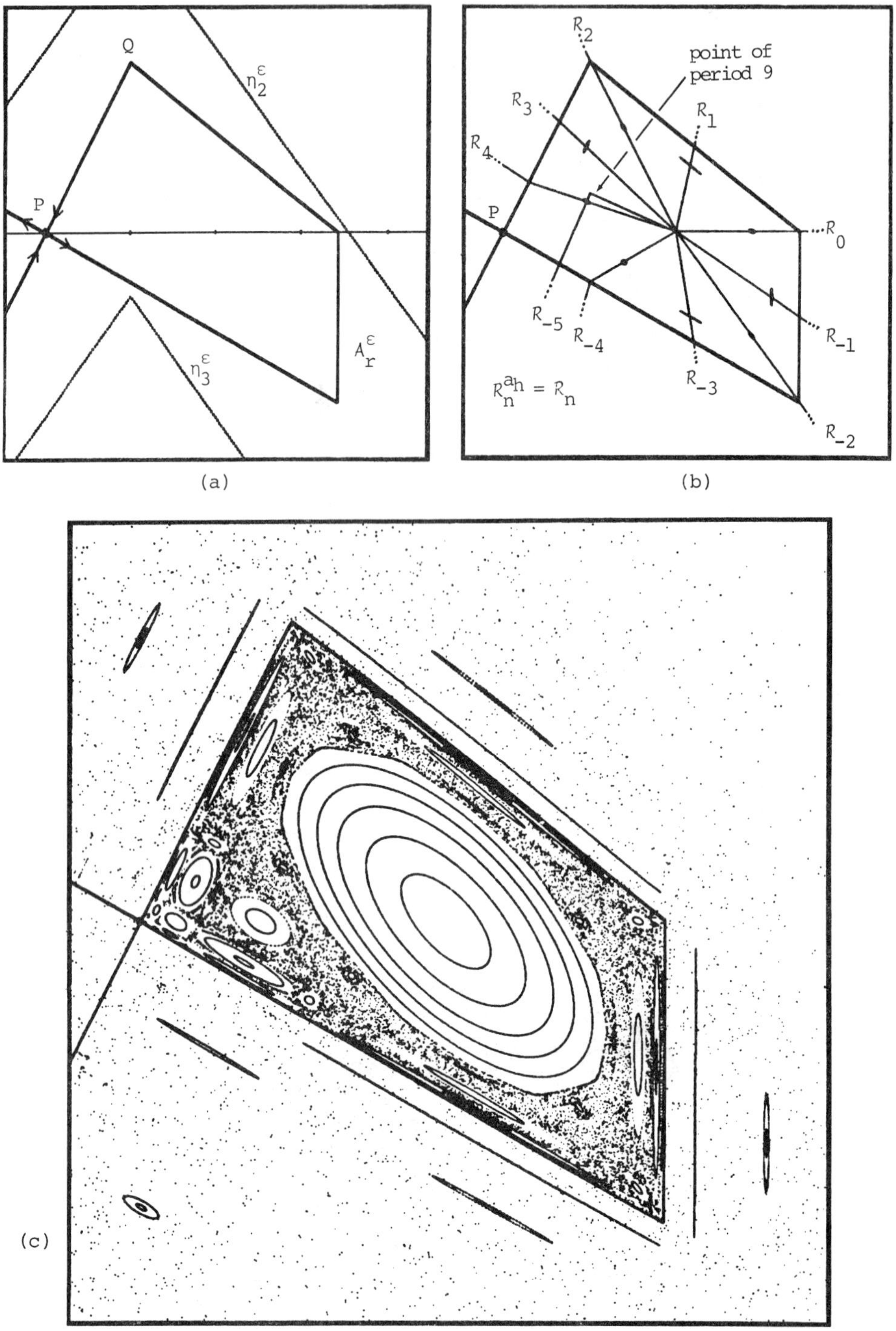

Figure 31. (singular separatrix , $h = \mu_2$)

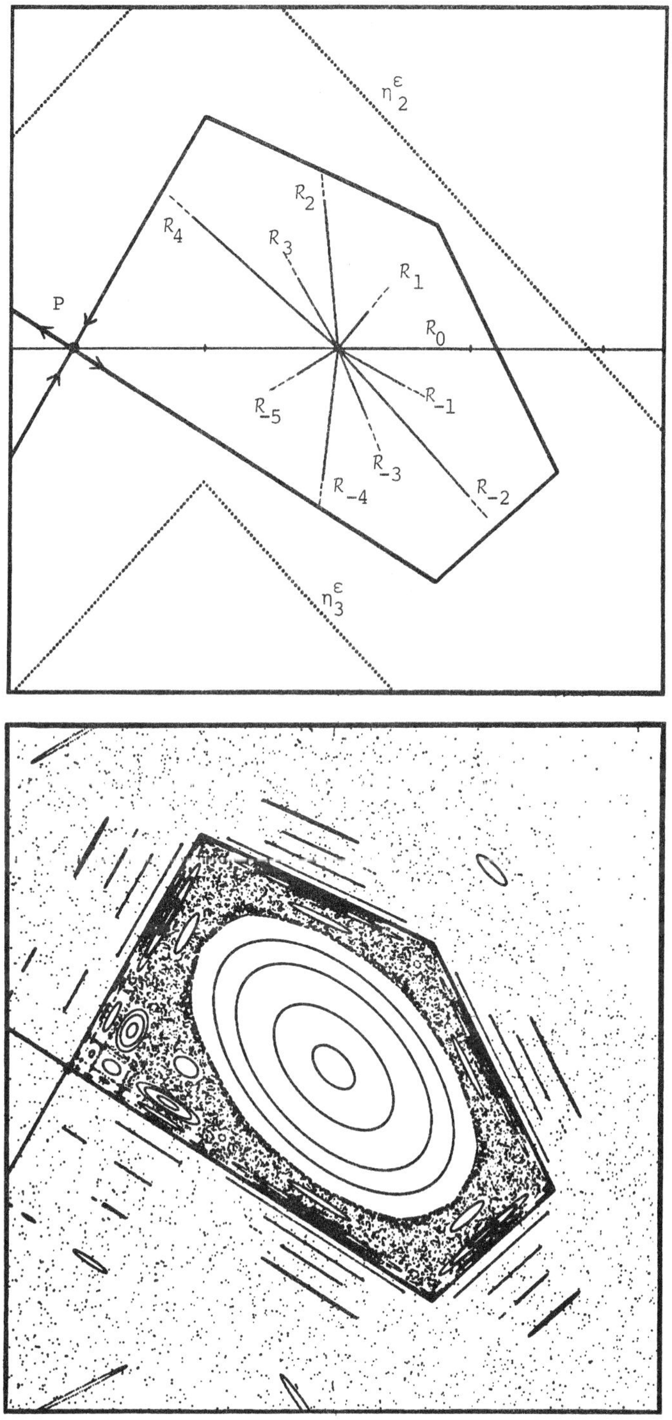

(a)

(b)

Figure 32. (singular separatrix , h = μ_3)

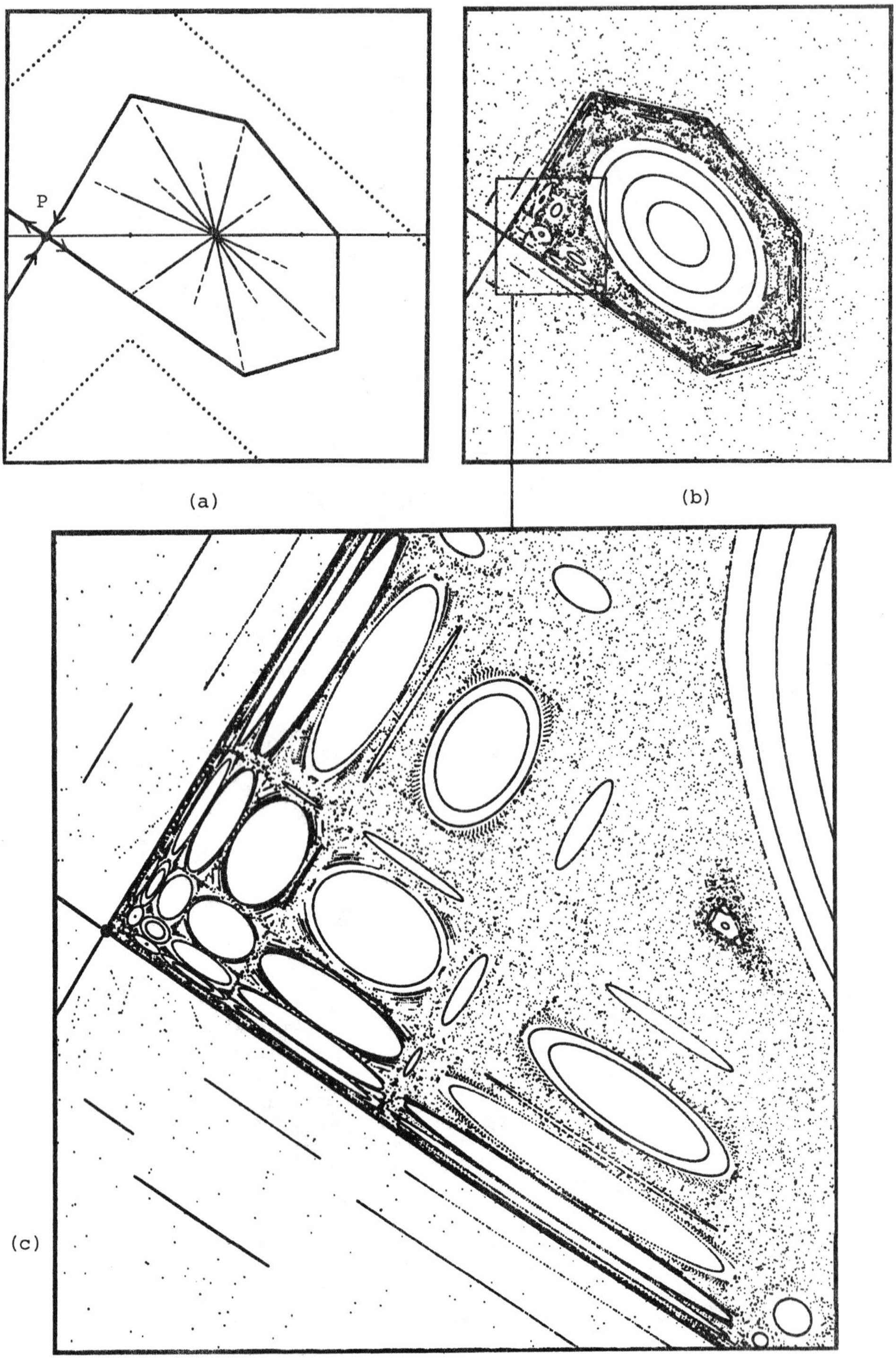

Figure 33. (singular separatrix , h = μ_4)

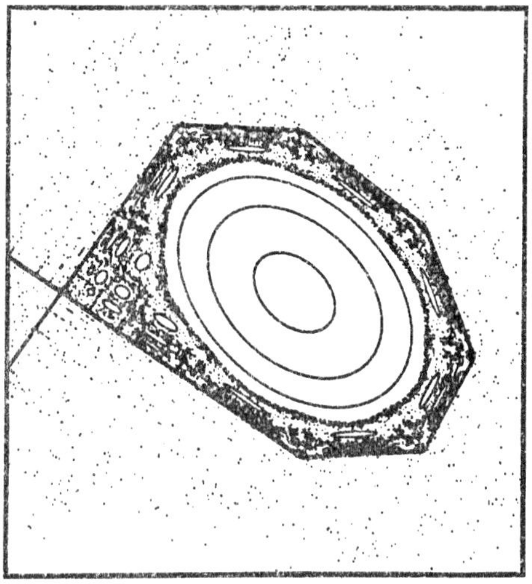

Figure 34. ($h = \mu_5$)

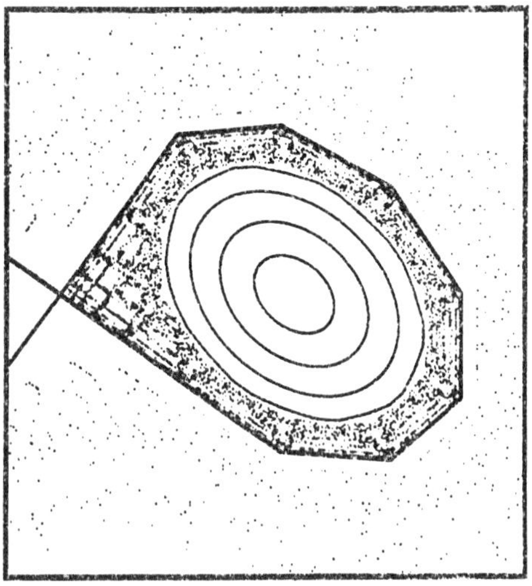

Figure 35. ($h = \mu_6$)

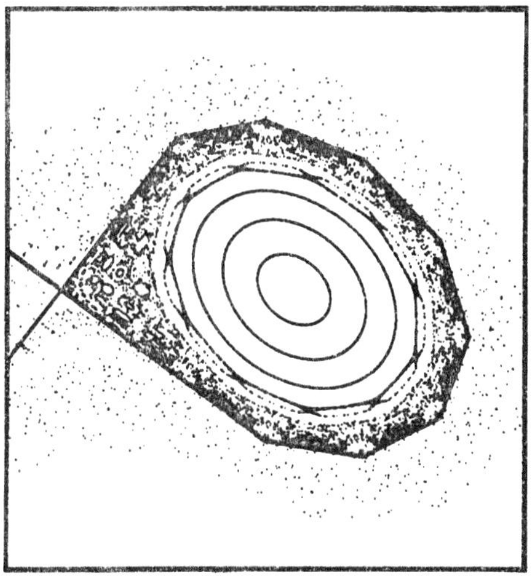

Figure 36. ($h = \mu_7$)

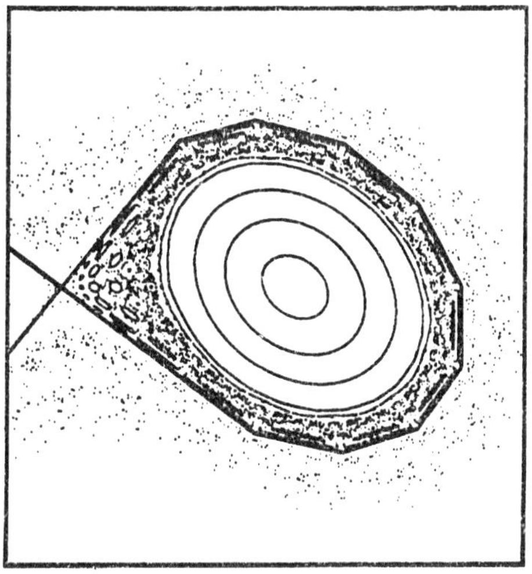

Figure 37. ($h = \mu_8$)

Then the y-component of $T^{-k}(Q_o^s)$ is decreasing and the x-component of $T^{-k}(Q_o^s)$ is increasing as k increases, provided $Q_{k-1}^s \in \bar{\mathcal{D}}$. One shows easily, for example, that $Q_1^s \in \bar{\mathcal{D}}$ for $h > \frac{1}{2}\sqrt{2}$ and $h \le \rho(\varepsilon)$ and $\varepsilon \in (0,2)$. For smaller values of h one can show that $Q_k^s(h) \in \bar{\mathcal{D}}$ for k sufficiently large and then follow the argumentation of the proof of theorem 1.2, using the affine linear structure on A_r^ε.

In figures (29+k), $k = 1, \ldots, 8$ we show experiments for $h = \mu_k$ and $\varepsilon \in (0,2)$ such that $\mu_k < \rho(\varepsilon)$. More precisely, in figure 30 we have $\varepsilon = 1.5$ and in figures (29+k), $k = 2, \ldots, 8$ we have $\varepsilon = 1$ in the admissible range according to theorem 1.3 (see also figure 38). In figure 38 we give a first (ε,h) phase diagram. For any pair (ε,h) such that $h = \mu_n$ and $\varepsilon \ge 2 + \mu_n - \sqrt{\mu_n^2 + 4}$ one has the total degeneration (1.38). Recall that for such choices of ε the y-component of Q_o^s is less or equal to 2. For choices of ε with $\varepsilon < 2 + \mu_n - \sqrt{\mu_n^2+4}$ the y-component of Q_o^s will exceed 2 and this has the consequence that $W_{out}^s(P;f_{\varepsilon,\delta})$, $\delta = 0$, will enter new domains of affine linearity of T.

$$\rho(\varepsilon) = (4\varepsilon - \varepsilon^2)/(4 - 2\varepsilon) = h$$

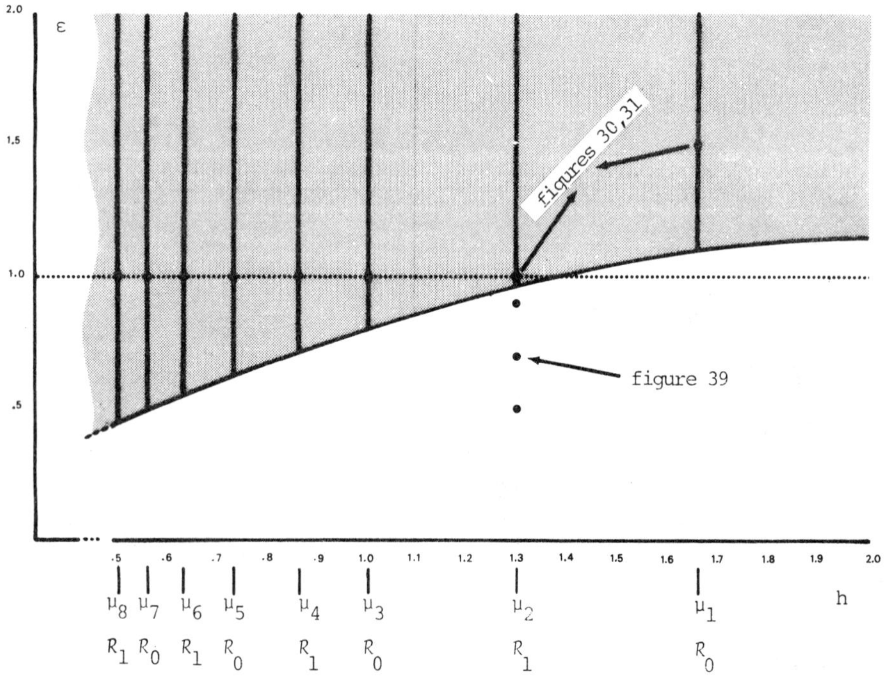

Figure 38. (phase diagram)

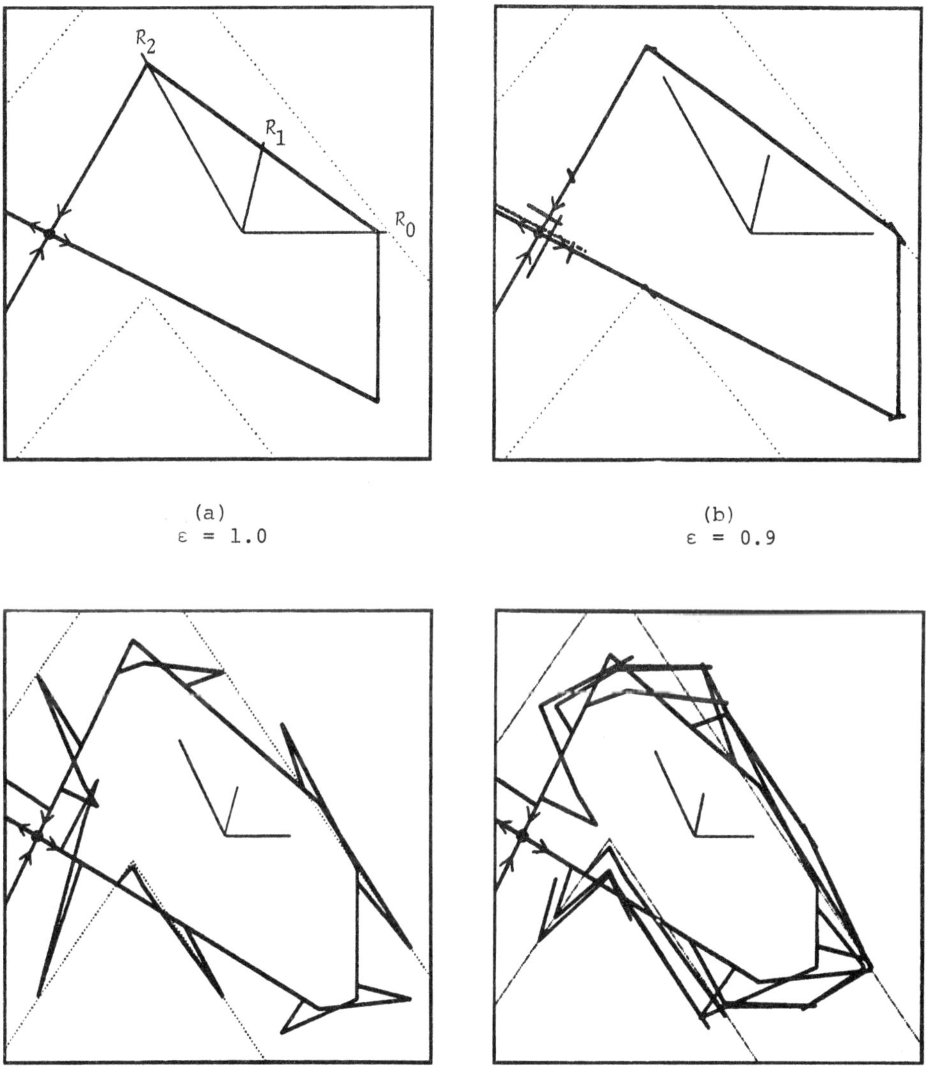

(a) $\varepsilon = 1.0$

(b) $\varepsilon = 0.9$

(c) $\varepsilon = 0.7$

(d) $\varepsilon = 0.5$

Figure 39. (degenerate homoclinic arcs)

Eventually, as ε is taken smaller and smaller (i.e. $f_{\varepsilon,\delta}$ is closer and closer to g_0) the complexity will increase drastically, because new domains of affine linearity of T come into play. The curve $\rho(\varepsilon) = h$ is just one of infinitely many curves in the (ε,h) phase diagram which give rise to phase transitions in the homoclinic structure. Figure 39 provides a series of computer experiments for $h = \mu_2$ and $\varepsilon = 1.0, 0.9, 0.7, 0.5$. One observes that the total degeneration (1.38) disappears for $h = \mu_2$ and $\mu_2 > \rho(\varepsilon)$. However, for $\varepsilon = 0.9, \ldots,$ one still has a degenerate homoclinic arc of odd type (see page 45 for notation). In our construction of totally degenerate homoclinic situations we have made an essential use of lemma 1.2 and the isochrones $R_n^{a_h}$. Figure 39 reveals that the partial degeneration for $\rho(\varepsilon) < \mu_2$ is in fact a degeneration at the isochrones $R_{2k+1}^{a_h}$, $k \in Z$. Careful experiments yield

<u>REMARK 1.5.</u> Let U be a small tubular neighborhood of $\varepsilon = h + 2 - \sqrt{h^2 + 4}$ in the (ε,h) phase diagram. Then

(i) For any $(\varepsilon,\mu_{2n-1}) \in U$ and $h = \mu_{2n-1}$, $n = 1, 2, 3, \ldots,$ $W^s_{out}(P;f_{\varepsilon,\delta})$ and $W^u_{out}(P;f_{\varepsilon,\delta})$, $\delta = 0$, have degenerate homoclinic arcs of odd type intersecting the isochrones $R_{2k}^{a_h}$ for all $k \in Z$.

(ii) For any $(\varepsilon,\mu_{2n}) \in U$ and $h = \mu_{2n}$, $n = 1, 2, 3, \ldots,$ $W^s_{out}(P;f_{\varepsilon,\delta})$ and $W^u_{out}(P;f_{\varepsilon,\delta})$, $\delta = 0$, have degenerate homoclinic arcs of odd type intersecting the isochrones $R_{2k+1}^{a_h}$ for all $k \in Z$.

Another remarkable aspect of the degenerations at $h = \mu_n$ is that they provide a stochastic dynamics which is confined to a bounded region. For each $h = \mu_n$ and $\mu_n < \rho(\varepsilon)$ $W^s_{out}(P;f_{\varepsilon,\delta})$, $\delta = 0$, separates the plane into two components which are invariant under T. For example in figure 33a we see a stochastic point structure which is obtained by several ten thousands of iterations of T based on a single initial value. The elliptic dynamics around the point $(8-2\varepsilon,0)$ is obvious, because $f_{\varepsilon,\delta}$, $\delta = 0$, is linear in a neighborhood of $8-2\varepsilon$. More surprising is the typical occurence of elliptic island structures in the stochastic point structure for each $h = \mu_n$. Figure 33c shows a magnification of a region around the hyperbolic fixed point $P = (4,0)$. This experiment supports the following

<u>CONJECTURE 1.4.</u> Let $\delta = 0$, $h = \mu_n$ and $\mu_n < \rho(\varepsilon)$. Then $W^s_{out}(P;f_{\varepsilon,\delta})$ separates the plane in a bounded domain I containing $(8-2\varepsilon,0)$ and unbounded domain O and there exist lattices $E^{in}_{ij} \subset I$, $E^{out}_{ij} \subset O$, $i \in N$, $j \in N$, such that

$P = (4,0)$ is a cluster point of E^{in}_{ij} and E^{out}_{ij} ;

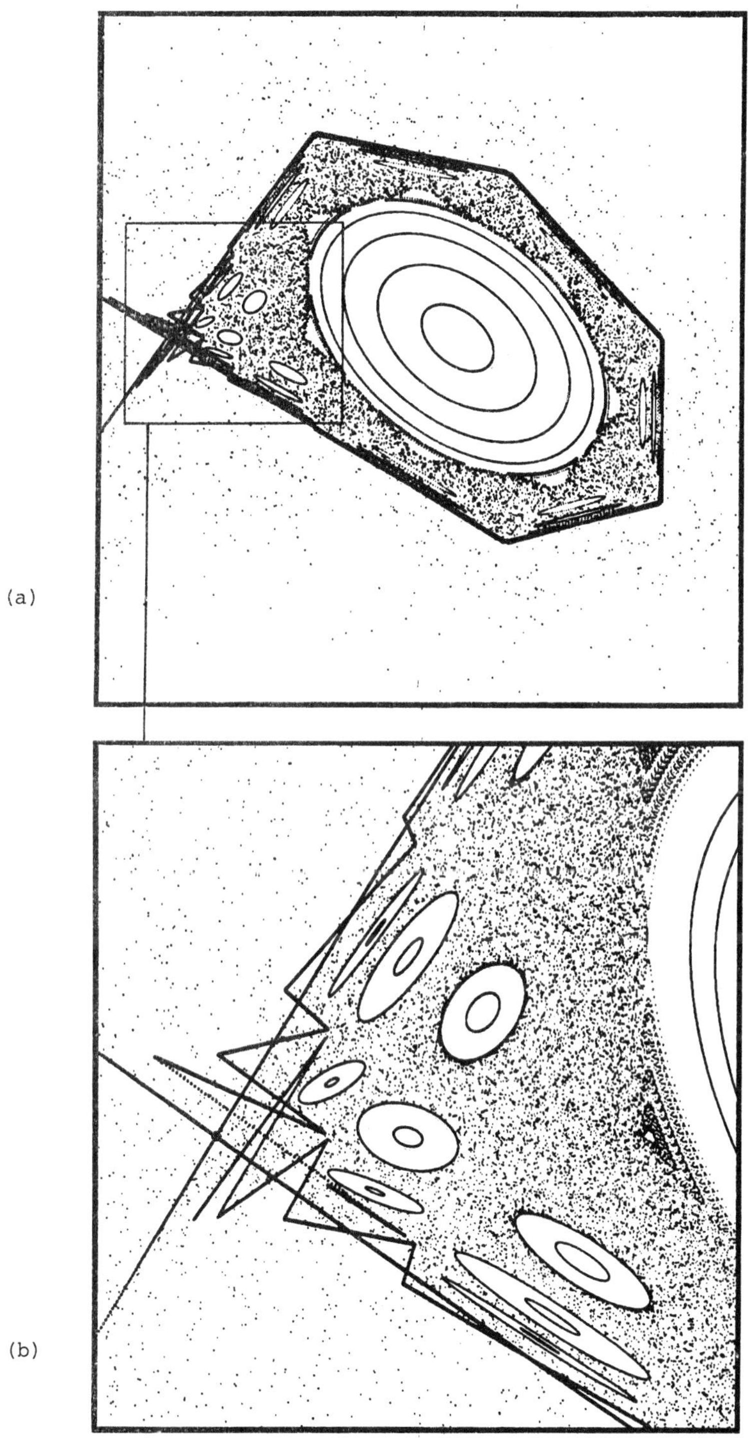

Figure 40. ($h = \mu_2 - 0.0036$, $\varepsilon = 1.0$)

-) let $X_k = W^s_{out}(P;f_{\varepsilon,\delta}) \cap \mathcal{R}^{a_h}_{2k+1}$, $k \in Z$, then the X_k are cluster points of E^{in}_{ij} and E^{out}_{ij} ;

-) each of the points E^{in}_{ij} and E^{out}_{ij} , $i \in N$, $j \in N$, is an elliptic periodic point.

This conjecture seems to contradict the general philosophy and experience which one usually has in connection with portraits of non-integrable diffeomorphisms. It can be shown by exploiting the affine linear structure of T together with lemma 1.3 . Note that, however, the situation in $h = \mu_n$ is non-generic. For $h = \mu_n \pm \bar{\varepsilon}$, $\bar{\varepsilon}$ small, all but finitely many of the elliptic structures in the stochastic point structure in I or O are destroyed and I and O loose their invariance properties. These properties are shown in figure 40, where $h = \mu_2 - 0.0036$ and $\varepsilon = 1$. Here the stochastic point structure is obtained by many different initial values. Iteration based on a single initial value eventually leaves the chosen window and tends not to return (at least for a reasonable number of iterations). Regarding the parameter h as a control parameter one observes significant transitions in the global dynamics as one manipulates h . There is a process of creation (h approaches a critical value μ_n) and destruction ($|h-\mu_n|$ grows) of elliptic structures and this process repeats an infinite number of times as $h \to 0$.

So far we have investigated the homo- and heteroclinic structures for the two-parameter family $T(f_{\varepsilon,\delta},h)$ $\varepsilon \in (0,2)$, $h > 0$, $\delta = 0$. In the final part we also vary $\delta > 0$, to discuss the critical values μ_n , $n = 1, 2, 3, \ldots,$ from a new point of view. This is the scenario of homoclinic bifurcation, which was introduced in [31] . Roughly speaking we mean by homoclinic bifurcation the following: Let $T = T(\mu)$ be an abstract one-parameter family of diffeomorphisms of the plane. Then $\mu = \mu_*$ is called a homoclinic bifurcation point, provided there is a creation (resp. destruction) of homoclinic points as μ passes through μ_* . There are tremendeous difficulties in making this first definition more precise. For example, given a transversal homoclinic point then due to the λ-lemma (see [25]) there exist infinitely many transversal homoclinic points in any neighborhood of the given one. Thus, it seems to be hopeless to study the fate of a single one when a parameter is changed. It turns out, however, (see [31]) that given symmetry relations as in lemma 1.2 that there is a natural definition of homoclinic bifurcation and by exploiting the symmetry structure one is able to prove (local) existence results of bifurcation and discuss (global) properties of the bifurcating objects (see [31,32]). We collect the following hypothesis for an abstract one-parameter family $T = T(\mu)$:

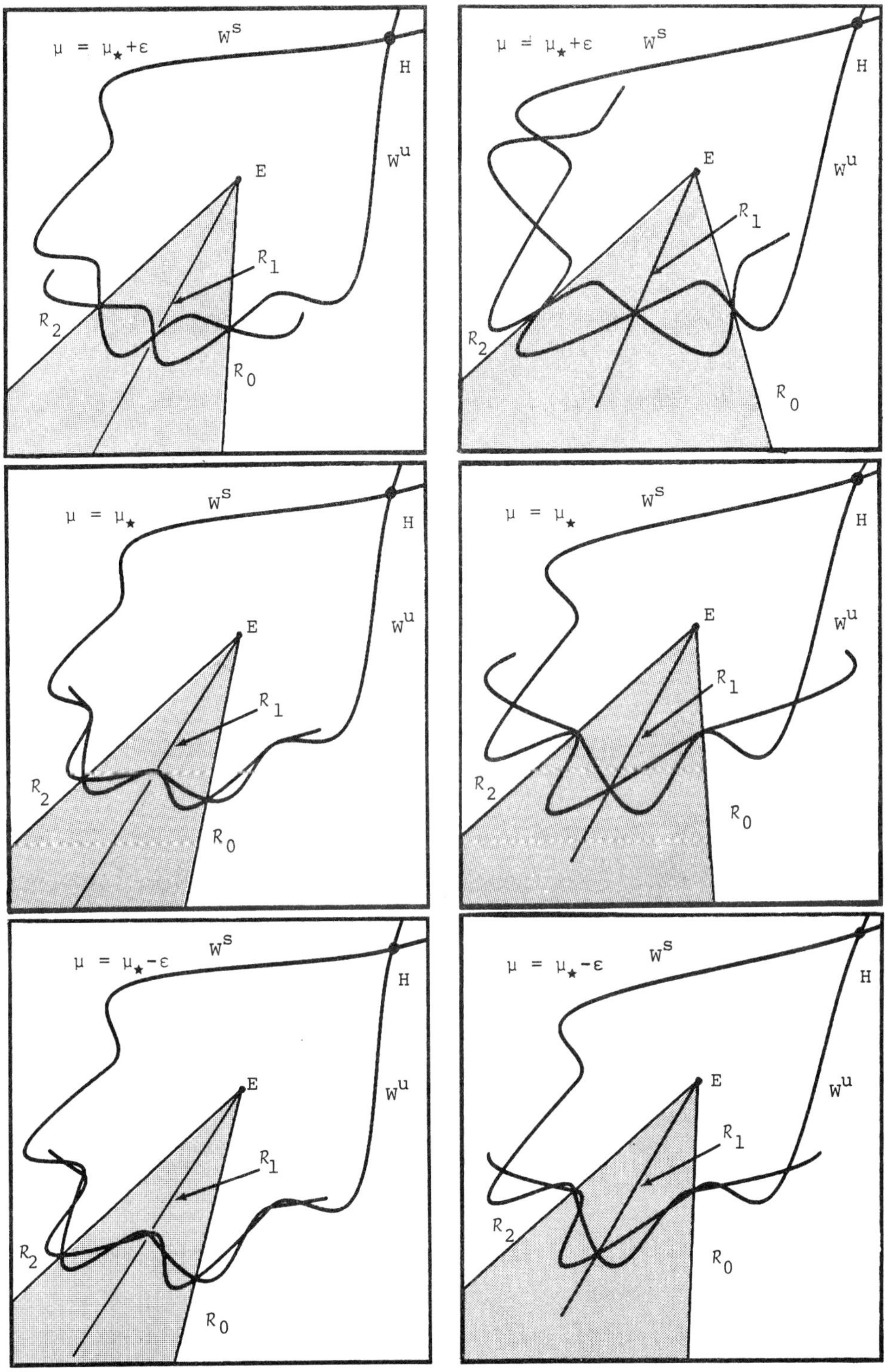

Figure 41. (homoclinic bifurcation at R_0 (right) and at R_1 (left))

(1.41) Let $T(\mu) = R(\mu)S(\mu)$ for all μ with $R^2(\mu) = S^2(\mu) = Id$. Let $T(\mu)(H) = H$ be a hyperbolic fixed point for all μ and let $T(\mu)(E) = \Xi$ be an elliptic fixed point for all μ. Let $\mathcal{R}_k(\mu) = \{z \in \mathbb{R}^2 : T^k(\mu)\, S(\mu)(z) = z\}$ and assume that $\mathcal{R}_o$ and $\mathcal{R}_1$ are unbounded and diffeomorphic to $\mathbb{R}$. Furthermore, let $\mathcal{R}_o \cap \mathcal{R}_2 = \{H,E\}$. Let $\mathcal{R}_o = r((-\infty, +\infty))$ with $r(-1)=H$, $r(0)=E$ and $\mathcal{R}_o^+ = r([0,+\infty))$. Let $\mathcal{R}_2^+ = T(\mathcal{R}_o^+)$. Then $\mathcal{R}_o^+ \cup \mathcal{R}_2^+$ separate the plane into two unbounded domains. Let C be the closed component which does not contain H. Assume that $(\mathcal{R}_1 \cap C) \setminus \{E\} \neq \emptyset$.

Note that C is a cone-like region with corner E (see figure 41). We will investigate the homoclinic structure only in C. Therefore let

$$\Psi^{s,u} : (-1, +1) \to \mathbb{R}^2$$

be immersions describing the invariant manifolds $W^{s,u}(H)$ with $\Psi^{s,u}(0) = H$. Now set

$$(1.42)\quad \begin{cases} t_1^{s,u} := \min\{t\in[0,1) : \Psi^{s,u}(t)\in C\} \\ t_2^{s,u} := \min\{t\in[0,1) : \Psi^{s,u}(t+\delta) \notin C \text{ and } \Psi^{s,u}(t-\delta) \in C \\ \qquad\qquad \text{for all } 0 < \delta << 1\} \\ W_C^{s,u}(H) = \begin{cases} \Psi^{s,u}([t_1^{s,u}, t_2^{s,u}]) \,, & \text{if } t_1^{s,u}\,,\ t_2^{s,u} \text{ exist,} \\ \emptyset \,, & \text{else.} \end{cases} \end{cases}$$

We can interpret $W_C^{s,u}(H)$ as the arcs of first intersection of $W^{s,u}(H)$ with C and can now split the homoclinic structure $W^s(H) \cap W^u(H)$ into a <u>primary</u> and <u>secondary</u> structure:

$$(1.43)\qquad \text{Prim}\,(H,\mu) = W_C^s(H) \cap W_C^u(H)\,.$$

With these preparations we define

<u>DEFINITION 1.4.</u> (see figure 41)
The critical parameter $\mu = \mu_*$ is called a point of <u>homoclinic bifurcation</u> at $\mathcal{R}_o$ (resp. $\mathcal{R}_1$) provided for all ε with $0 < \varepsilon << 1$:

$$\emptyset \neq \text{Prim}\,(H,\mu_*-\varepsilon) \subset R_o \cup R_1 \cup R_2 ,$$

$$\emptyset \neq \text{Prim}\,(H,\mu_*+\varepsilon) \not\subset R_o \cup R_1 \cup R_2 ,$$

Prim $(H,\mu_*) \cap R_o$ (resp. Prim $(H,\mu_*) \cap R_1$) is a degenerate homoclinic point of odd type and if $\text{Bif}(H,\mu_*+\varepsilon) = \text{Prim}\,(H,\mu_*+\varepsilon) \setminus (R_o \cup R_1 \cup R_2)$ then

$$\text{Bif}(H,\mu_*+\varepsilon) \to \text{Prim}(H,\mu_*) \cap R_o$$
$$(\text{resp. Bif}(H,\mu_*+\varepsilon) \to \text{Prim}(H,\mu_*) \cap R_1)$$

as $\varepsilon \to 0$.

Thus, the symmetry structure allows us to distinguish bifurcating homoclinic points. With the Birkhoff-Smale theorem (see [38] , p. 775) in mind one expects that homoclinic bifurcation should initiate an infinite series of bifurcations for periodic points:

PROPOSITION 1.5. Let $T = T(\mu)$ be a one-parameter family of diffeomorphisms with $T(\mu) = R(\mu) \circ S(\mu)$ for all μ and $R^2(\mu) = S^2(\mu) = \text{Id}$. Assume that (1.41) is satisfied and assume that $\mu = \mu_*$ is a point of homoclinic bifurcation at R_o (resp. R_1) . Furthermore, assume that

$$\text{Prim}(H,\mu_*\pm\varepsilon) , \quad 0 < \varepsilon << 1 ,$$

is transversal and that $W_C^{s,u}(H)$ intersect R_o and R_1 transversally for any $\mu = \mu_* + \varepsilon$, $0 \leq |\varepsilon| << 1$. Then there exists a family $\{P_k(\mu)\}_{k\in N}$ and a sequence $\{\mu_k\}_{k\in N}$ such that

- $P_k(\mu)$ is a periodic point of $T(\mu)$ with period p_k ,
- $\mu_k \to \mu_*$ as $k \to \infty$,
- $p_k \to \infty$ as $k \to \infty$,
- $p_k(\mu) \in R_o$ (resp. R_1)
- the family $p_k(\mu)$ (k fixed) undergoes a bifurcation at $\mu = \mu_k$.

PROOF. The proof is essentially an immediate consequence of the λ-lemma and our definition. For example, let μ_* be a point of bifurcation at R_o .

Then there is a family $P_\infty(\mu)$, $\mu \in (\mu_*-\varepsilon,\mu_*+\varepsilon)$ of homoclinic points $P_\infty(\mu) \in R_o$, which is transversal for $\mu \neq \mu_*$. By assumption, R_o intersects $W^s_C(H)$ and $W^u_C(H)$ transversally for any $\mu \in (\mu_*-\varepsilon,\mu_*+\varepsilon)$. Thus, there exists a small 1-cell I_0 on R_o such that

$$R_{2k}(\mu) \supset T^k(\mu)(I_0) \text{ is arbitrarily } C^1\text{-close to } W^u_C(H) \text{ , and}$$

$$R_{-2k}(\mu) \supset T^{-k}(\mu)(I_0) \text{ is arbitrarily } C^1\text{-close to } W^s_C(H) \text{ ,}$$

for $\mu \in (\mu_*-\varepsilon , \mu_*+\varepsilon)$ and k sufficiently large. In particular, $T^k(\mu)(I_0) \cap I_0 \neq \emptyset$ and we choose $P_k(\mu) \in T^k(\mu)(I_0) \cap I_0$. Then $P_k(\mu)$ is a point of period $p_k = k$ and $P_k(\mu) \in T^{-k}(\mu)(I_0) \cap I_0$. Since $\mathrm{Prim}(H,\mu_*\pm\varepsilon)$ is transversal we may conclude that for ε fixed but sufficiently small, and $k \in N$ sufficiently large

$$\{T^k(\mu)(I_0) \cap T^{-k}(\mu)(I_0)\} \setminus \{R_o \cup R_1 \cup R_2\}$$

is non-empty for $\mu = \mu_* + \varepsilon$ and C^o-close to $\mathrm{Bif}(H,\mu_*+\varepsilon)$, while the above set is empty for $\mu = \mu_* - \varepsilon$. Thus, $P_k(\mu)$ undergoes a bifurcation for $\mu = \mu_k \in (\mu_*-\varepsilon,\mu_*+\varepsilon)$.

The question arises whether one is able to verify the assumptions of proposition 1.5 in a concrete example. This has been shown in [31,32],where

$$T(\mu)\begin{pmatrix} x \\ y \end{pmatrix} = \begin{pmatrix} 2x-y-\mu h_\delta(x) \\ x \end{pmatrix} ,$$

for a generating nonlinearity h_δ in the spirit of $f_{\varepsilon,\delta}$, where

$$h_\delta(s) = \begin{cases} s & , \quad s \leq 0 \\ s_1-s & , \quad s \geq \delta \\ \phi_\delta(s) & , \quad 0 \leq s \leq \delta , \end{cases}$$

and $s_1 > 0$, $0 < \delta < s_1$, and ϕ_δ is a C^∞-function which also makes h_δ a C^∞-function, and $h_\delta(s) > 0$ for $0 < s < s_1$. Figure 42 gives a phase-diagram for this particular (δ,μ)-family, when ϕ_δ is a linear function. We see a partition of the (δ,μ)-plane into shaded and white regions. Any path $\gamma(t)$ in that plane fixes a one-parameter family $T(\gamma(t))$. If $\gamma(t)$ traverses a shaded region then the one-parameter family exhibits homoclinic bifurcation as $\gamma(t)$ enters or leaves a shaded region. Points in the phase-diagram where shaded regions hit together in a singular point correspond to totally degenerate homoclinic structures in the sense of

(1.38).

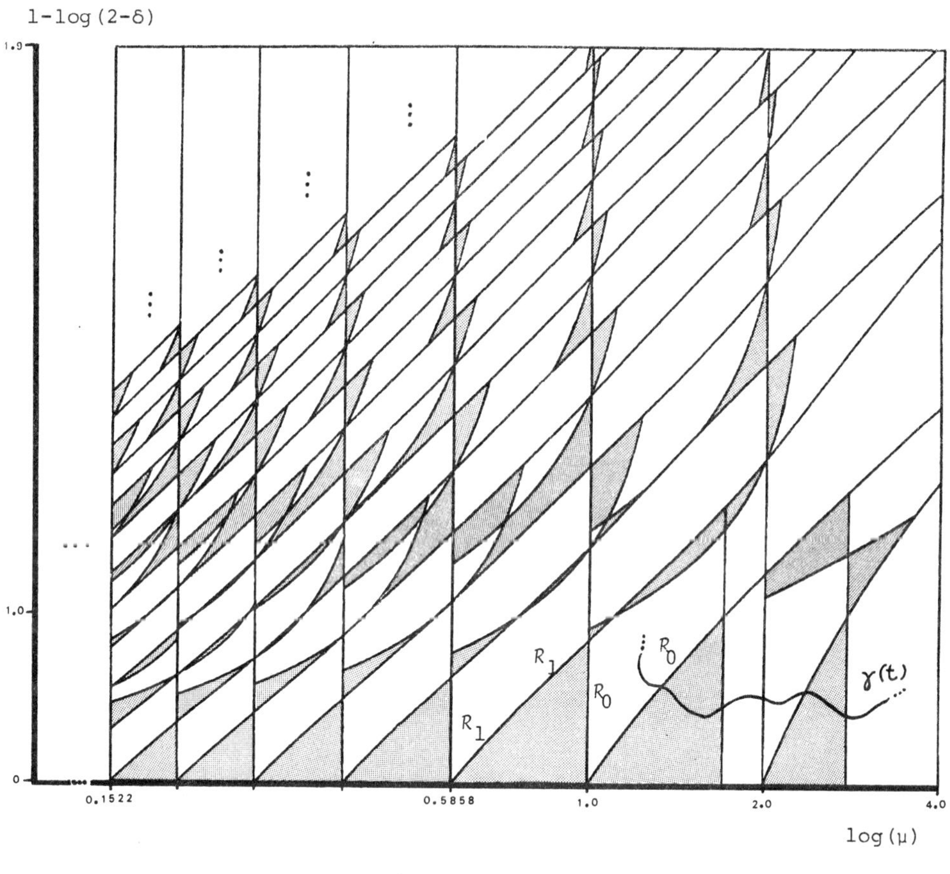

Figure 42. (phase diagram of homoclinic bifurcation)

In principle one could carry out the samme discussion as in [31,32] for the models of T which are presented in this paper. For reasons of length we restrict ourselves to a short presentation of careful numerical experiments, which suggest that our generating nonlinearity $f_{\varepsilon,\delta}$ (see example 1.2) will be adequately described by a phase-diagram with features similar to those shown in figure 42. In this view figure 38 is a first step. We choose $f_{\varepsilon,\delta}$ according to example 1.2, set $\varepsilon = 1$ and study our model of $T(f_{\varepsilon,\delta},h)$ generated by $f_{\varepsilon,\delta}$ according to (1.6) in a small neighborhood of $(\mu_2,0)$ in (μ,δ)-space, μ_2 as in theorem 1.3 :

$$(h,\delta) \in (\mu_2 - \bar{\varepsilon}\ ,\ \mu_2 + \bar{\varepsilon}) \times (-\bar{\delta}\ ,\ \bar{\delta})\ ,$$

where $\bar{\varepsilon} = 5\cdot 10^{-3}$, $\bar{\delta} = 5\cdot 10^{-3}$.

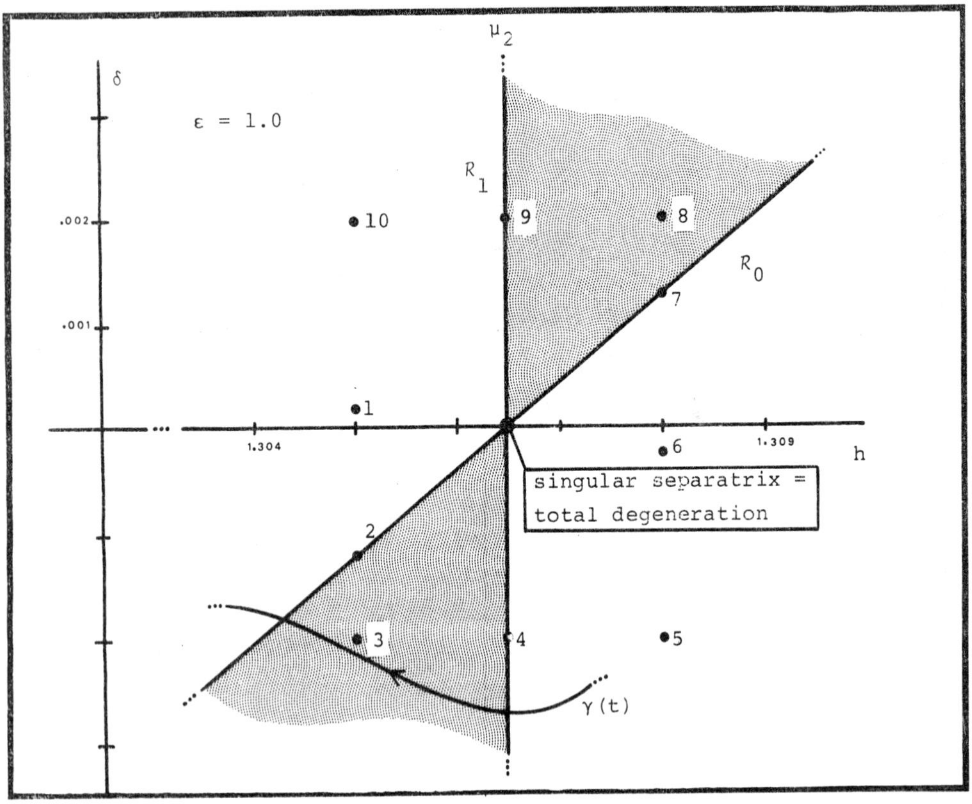

Figure 43. (local phase diagram of homoclinic bifurcation)

Figure 43 shows a shaded area in the (h,δ) parameter space. Let $\gamma : R \to R^2$ be a section (see figure 43), which traverses the shaded region. Then there is homoclinic bifurcation at R_{2k+1} for the t-family $T(f_{\varepsilon,\delta}, \gamma(t))$ at $\gamma(t) = \mu_2$, $\mu_2 = 1.3065\ldots$, and there is homoclinic bifurcation at R_{2k} , where $\gamma(t)$ leaves the shaded area again. For parameter values (h,δ) in the shaded region one has homoclinic points in $\mathrm{Prim}(P,(h,\delta))$ which are not in $R_o^{a_h} \cup R_1^{a_h} \cup R_2^{a_h}$ whereas for parameter values (h,δ) outside the shaded regions one has $\mathrm{Prim}\,(P,(h,\delta)) \subset R_o^{a_h} \cup R_1^{a_h} \cup R_2^{a_h}$; $P = (4,0)$. To make this visible we show 10 choices of parameters (h,δ) in a small neighborhood of

$(h,\delta) = (\mu_2, 0)$, all for $\varepsilon = 1$, as indicated in figure 44. Figure 45 shows magnifications of $T^2(C)$ (see figure 44); i.e. rather than studying the homoclinic structure of $W^{s,u}_{out}(P, f_{\varepsilon,\delta})$, $\varepsilon = 1$, in C we study it equivalently in the cone-like region $T^2(C)$, which is bounded by R_4 and R_6 .

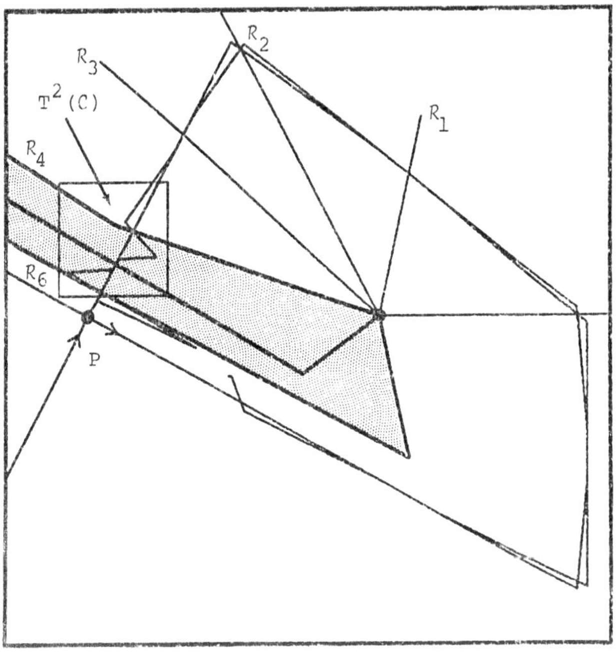

Figure 44. (homoclinic structure and iterated 'cone' near P=(4,0))

<u>REMARK 1.6</u>. Note that according to theorem 1.3 we have a total degeneration for $\varepsilon = 1$, $\delta = 0$ and $h = \mu_2$. Thus, figure 45 shows the change of the primary homoclinic structure as δ and h vary close to that singularity and $\varepsilon = 1$. According to figure 38 one should expect similar structures near the other degenerations $h = \mu_k$, $k = 3, 4, \ldots$. This is indeed supported by numerical experiments. Thus, we may conjecture that the critical parameters $h = \mu_k$ (see theorem 1.3) , which are total degenerations for the model $T(f_{\varepsilon,\delta}, h)$, $\delta = 0$, $h < \rho(\varepsilon)$, are in fact points of homoclinic bifurcation for the model $T(f_{\varepsilon,\delta}, h)$, $\delta \neq 0$. In this view we can interpret the total degenerations for $h = \mu_k$, $h < \rho(\varepsilon)$, $\delta = 0$ as limiting cases of homoclinic bifurcation (see figure 43) .

We conclude with numerical experiments for C^∞-models of T . One may

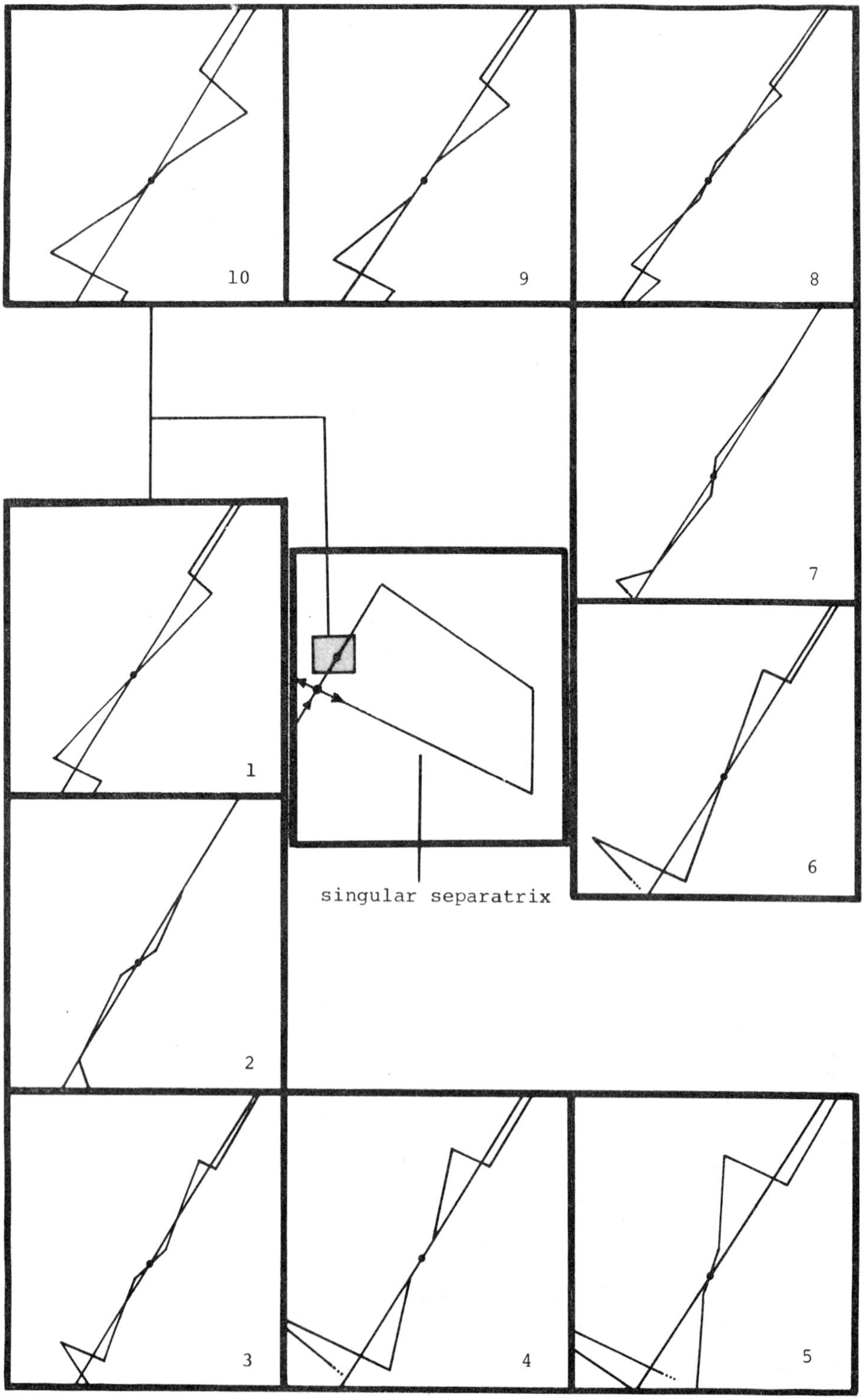

Figure 45. (10 experiments for $T(f_{\varepsilon,\delta},h)$ according to figure 43)

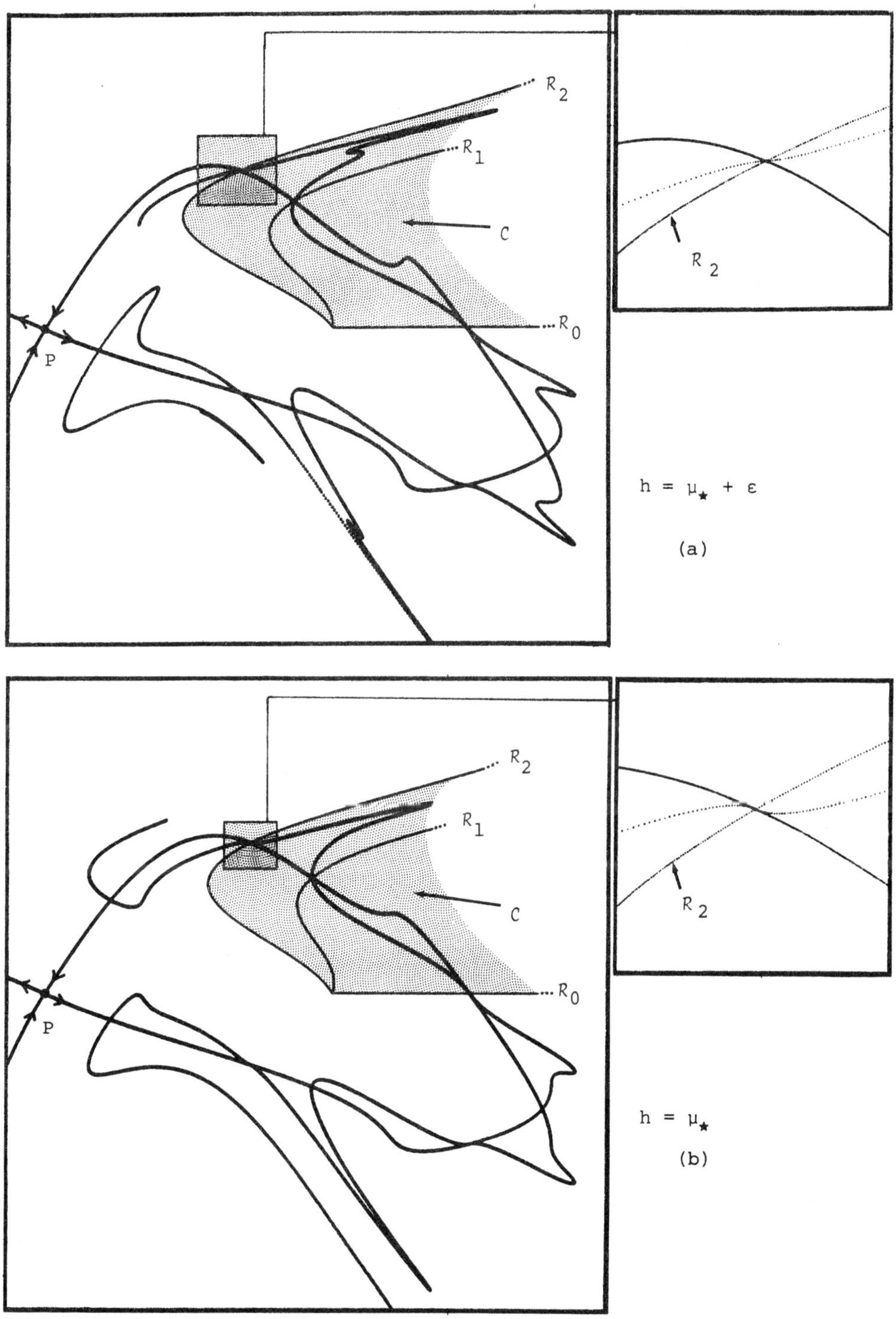

Figure 46. (homoclinic bifurcation for smooth model $T(f_\varepsilon, h)$)

suspect that the above phenomena of homoclinic bifurcation are an artefact of PL-models. This is not the case and this is demonstrated in figure 47, where the generating nonlinearity is f_ε from example 1.1 with $\varepsilon = 1.0$. Figure 46 (a-c) shows bifurcation at $R_{2k}^{a_h}$ for $h = \mu_* \approx 1.26$ and the cones C. Figure 47 is an illustration of proposition 1.5 at that point of homoclinic bifurcation. It shows how $R_k^{a_h}$ and $R_{-k}^{a_h}$ mimic the bifurcation in $W_{out}^s(P,f_\varepsilon)$, $W_{out}^u(P,f_\varepsilon)$ and initiate a period doubling bifurcation.

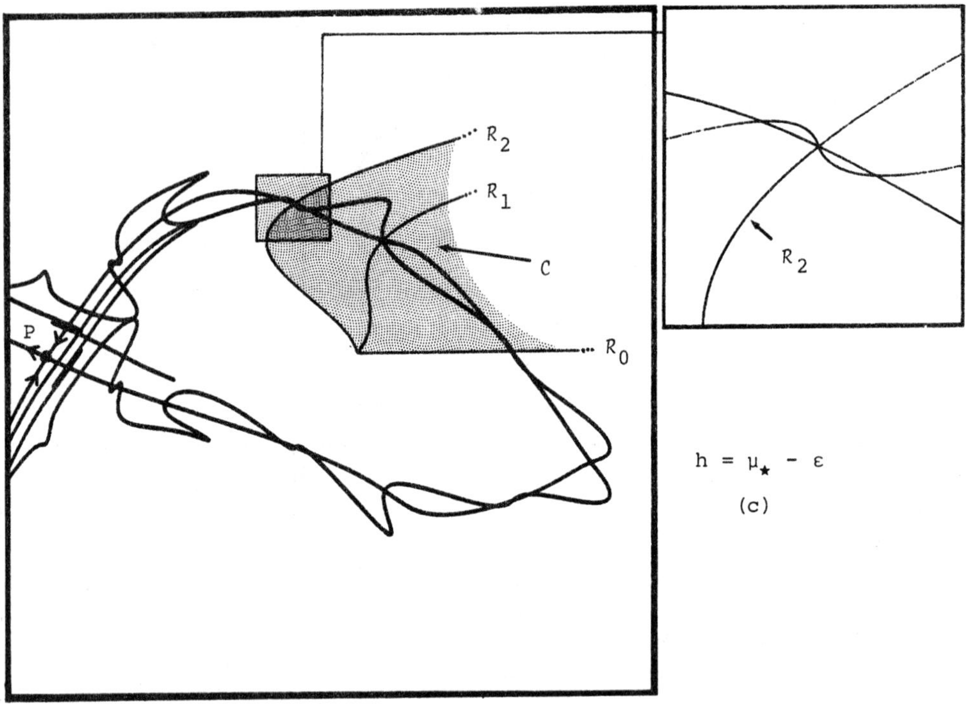

$h = \mu_\star - \varepsilon$

(c)

Figure 46. (continued)

The isochrones $R_{2k+1}^{a_h}$ in the above figures are obtained (see lemma 1.3) by

$$R_{2k-1}^{a_h} = T^k(R_{-1}^{a_h}) , \quad k \in Z ,$$

where

$$R_{-1}^{a_h}(h) = \{(x,y) : y = - {}^h/_2 f(x)\}$$

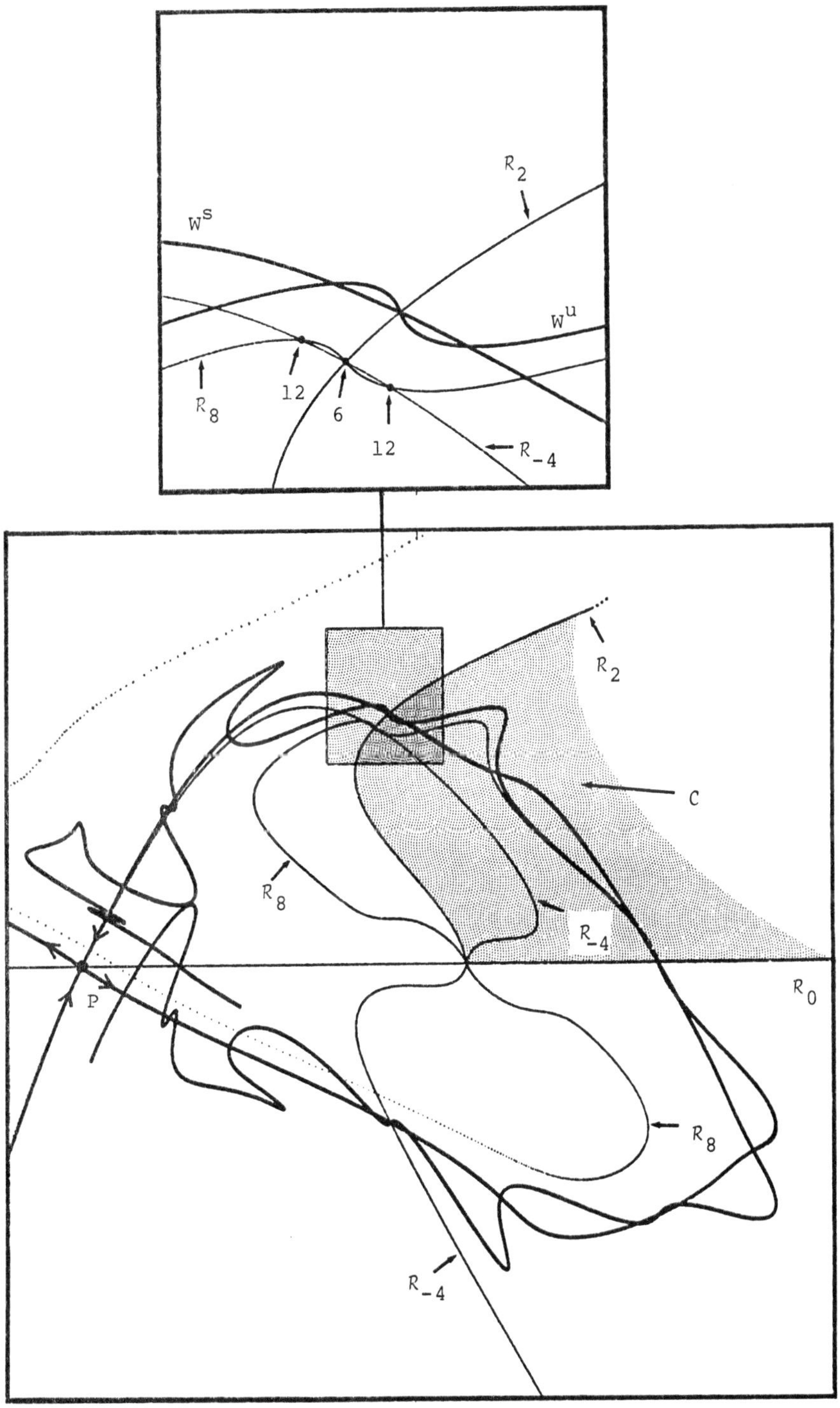

Figure 47. (bifurcation of periodic points near homoclinic bifurcation)

for any generating function f. This follows, because (see (1.8) and (1.9))

$$T^{-1}a_h = a_hT = -b_ha_hb_h .$$

All computer plots and numerical experiments were done in the laboratory of the second author at the "Forschungsschwerpunkt Dynamische Systeme, Universität Bremen". The experiments were made possible by a very elegant and most helpful general purpose interactive graphics-software developed by H. Jürgens, which is designed for experimental studies of the time-discrete dynamics of diffeomorphisms of the plane.

CHAPTER 2

2. A PRIORI BOUNDS AND EXISTENCE OF PERIODIC SOLUTIONS.

We shall study in this section the problem of *a priori* bounds and existence for periodic solutions of

$$(2.1) \qquad \dot{x}(t) = -\alpha f(x(t-1))$$

Following the standard terminology, a periodic solution of (2.1) will be called "slowly oscillating" if there exist positive numbers $z_1 > 1$ and $z_2 > z_1 + 1$ such that $x(0) = 0$, $x(t) > 0$ for $0 < t < z_1$, $x(t) < 0$ for $z_1 < t < z_2$ and $x(t+z_2) = x(t)$ for all t.

Aside from some results in another section, we shall be exclusively concerned with the case that $f(x)$ is odd, but unlike Kaplan and Yorke [19], we shall not assume that $f(x) > 0$ for all $x > 0$. The fact that $f(x)$ may change sign for $x > 0$ leads to striking differences from the theory in [19]. As in [29], we shall call $x(t)$ a "special periodic solution" of (1.1) if (1) $x(-t) = -x(t)$ for all t, (2) $x(t+2) = -x(t)$ for all t, and $x(t)$ is not identically zero. We shall call a special periodic solution "positive" if $x(t) \geq 0$ for $0 \leq t \leq 2$.

Kaplan and Yorke observed in [19] that if $f(x)$ is positive for $x > 0$ and an odd function and if $x(t)$ is a special periodic solution of (2.1) and $y(t) = x(t-1)$, then $(x(t),y(t))$ satisfies the O.D.E.

$$(2.2) \qquad \begin{cases} \dot{x}(t) = -\alpha f(y(t)) \\ \dot{y}(t) = \alpha f(x(t)) \end{cases}$$

Conversely, suppose that $(x(t),y(t))$ satisfies (2.2) and that $(x(0),y(0)) = (0,-c)$ for some $c > 0$. They then prove that $(x(t),y(t))$ is periodic and that if $T = T(c)$ is its minimal period and $r = (\frac{1}{4})T$, then $y(t) = x(t-r)$, $x(t+2r) = -x(t)$, $x(-t) = -x(t)$ and $x(t)$ is positive for $0 < t < 2r$. In particular $x(t)$ satisfies the differential-delay equation

$$(2.3) \qquad \dot{x}(t) = -\alpha f(x(t-r))$$

Our first lemma considers how many of the Kaplan-Yorke results carry over to our class of functions. The major problems are that orbits of (2.2) may start on the y-axis and never intersect the x-axis and that orbits may circle the origin in a clockwise direction (only counter-clockwise is possible in the Kaplan-Yorke case).

LEMMA 2.1. Let $f: R \to R$ be an odd, locally Lipschitzian function and suppose that $(x(t),y(t))$ is a solution of (2.2) such that $(x(0),y(0)) = (0,-c)$, $c > 0$. Assume that there exists a time $t > 0$ such that $y(t) = 0$ and define $r = \inf\{t>0: y(t)=0\}$. Then it follows that $x(t) \neq 0$ for $0 < t < 2r$, $x(t+2r) = -x(t)$ for all t and $x(-t) = -x(t)$ for all t. If $x(r) > 0$, $x(t)$ satisfies equation (2.3), while if $x(r) < 0$, $x(t)$ satisfies

$$(2.4) \qquad \dot{x}(t) = \alpha f(x(t-r))$$

PROOF. Assume first that $x(r) = d > 0$; we will discuss the case $x(r) < 0$ later. Notice that $x(r) \neq 0$, because $(0,0)$ is a rest point of (2.2). We will prove later that $x(t) > 0$ for $0 < t \leq r$, so assume this fact for the moment.

Under the above assumptions, the curve $(x(t),y(t))$ must cross the line $y = -x$, so there exists a time τ such that $0 < \tau < r$ and

$$(2.5) \qquad x(\tau) = -y(\tau) > 0$$

If we define $(x_0(t),y_0(t)) = (x(\tau+t),y(\tau+t))$ and $(x_1(t),y_1(t)) = (-y(\tau-t),-x(\tau-t))$, a calculation (using the oddness of f) shows that $(x_0(t),y_0(t))$ and $(x_1(t),y_1(t))$ both satisfy (2.1) and have the same initial value, so for all t

$$(2.6) \qquad \begin{cases} x(\tau+t) = -y(\tau-t) \\ y(\tau+t) = -x(\tau-t) \end{cases}$$

Equation (2.6) implies that $y(t) < 0$ for $0 \leq t < 2\tau$ and that $y(2\tau) = 0$. The minimality of r then implies that $r = 2\tau$ and

$$(2.7) \qquad x(r) = x(2\tau) = -y(0) = c$$

If we define $(x_2(t),y_2(t)) = (y(t+r), -x(t+r))$, a calculation shows that $(x_2(t),y_2(t))$ satisfies (2.2) and has the same initial value as $(x(t),y(t))$, so for all t

$$(2.8) \qquad \begin{cases} y(t+r) = x(t) \\ -x(t+r) = y(t) \end{cases}$$

Equation (2.8) immediately implies that for all t

$$(2.9) \qquad \begin{cases} x(t+2r) = -x(t) \\ y(t) = x(t-r) \end{cases}$$

and also that $x(t) > 0$ for $0 < t < 2r$.

To see that $x(-t) = -x(t)$, observe that if

$$(x_3(t), y_3(t)) := (-x(-t), y(-t))$$

then $(x_3(t), y_3(t))$ satisfies (2.2) and has the same initial value as $(x(t), y(t))$ at $t = 0$.

It remains to show that $x(t) > 0$ for $0 < t \le r$ (assuming $x(r) > 0$) . Define a number σ by

$$\sigma = \sup\{t\colon 0 \le t < r\ ,\ \ x(t) = 0\}$$

We wish to show that $\sigma = 0$. If we define

$$(x_4(t), y_4(t)) = (x(t+\sigma), y(t+\sigma))$$

then $y_4(0) < 0$, $x_4(t) > 0$ for $0 < t \le r_1 = r - \sigma$ and (x_4, y_4) satisfies equation (2.2). It follows from the work we have already done that

$$(2.10)\qquad \begin{cases} x_4(t) > 0 \quad \text{for} \quad 0 < t < 2r_1 \\ x_4(t) = -x_4(-t) \\ y_4(t+2r_1) = -y_4(t) \end{cases}$$

If $0 < \sigma < 2r_1$ equation (2.10) implies that

$$-x_4(\sigma) = x(0) < 0$$

which is a contradiction. Therefore, we assume that $\sigma \ge 2r_1$ or $\sigma = 0$. Equation (2.10) implies that

$$(2.11)\qquad y(t+2r_1) = -y(t)$$

for all t . If $\sigma \ge 2r_1$, we know that $2r_1 < r$; equation (2.11) implies that $y(2r_1) > 0$, so $y(t)$ has a zero on $[0, 2r_1]$, which contradicts the minimality of r . We conclude that $\sigma = 0$.

We still must consider the case $x(r) < 0$. Exactly as in the other case we prove that $x(t)$ is negative for $0 < t < 2r$, $x(t+2r) = -x(t)$ and $x(-t) = -x(t)$ for all t . However, now we have $x(r) = y(0)$, so if we define $(x_5(t), y_5(t))$ by

$$\begin{cases} x_5(t) = -y(t+r) \\ y_5(t) = x(t+r) \end{cases}$$

$(x_5(t), y_5(t))$ satisfies (2.2) and has the same initial value as $(x(t), y(t))$. We find, therefore, that for all t

$$(2.12) \qquad \begin{cases} x(t) = -y(t+r) \\ y(t) = x(t+r) \end{cases}$$

It follows easily from (2.12) and (2.2) that $x(t)$ satisfies (2.4).

REMARK 2.1. Kaplan and Yorke [19] observe that if $\varphi(x) := \int_0^x f(s)ds$, then any solution of (2.2) satisfies

$$(2.13) \qquad \varphi(x(t)) + \varphi(y(t)) = \text{constant}.$$

If $x\varphi(x) > 0$ for all nonzero x, most of Lemma 2.1 follows immediately from equation (2.13). In our case, more care is needed.

It will be convenient to recall some notation and results from [29]. Define a Banach space X by

$$(2.14) \qquad X = \{x: R \to R \mid x(t+2) = -x(t) \text{ and } x(-t) = -x(t) \text{ for all } t \text{ and } x \text{ is continuous}\}$$

If $x \in X$, $\|x\| = \sup_t |x(t)|$. Also define a cone $K \subset X$ by

$$(2.15) \qquad K = \{x \in X: x(t) \geq 0 \text{ for } 0 \leq t \leq 2\}$$

If $f: R \to R$ is odd and continuous, and if an operator F is defined by

$$(2.16) \qquad (Fx)(t) = -\int_0^t f(x(s-1))ds$$

then it is a special case of results in [29] that F is a continuous, compact map of X into itself. Finding special periodic solutions of (2.1) is equivalent to finding $x \in X - \{0\}$ and $\alpha \in R$ such that

$$(2.17) \qquad x = \alpha F(x).$$

If $f'(0) = 1$, it is proved in [29] that F is Fréchet differentiable at 0 with Fréchet derivative $L: X \to X$ given by

$$(2.18) \qquad (Lx)(t) = -\int_0^t x(s-1)ds.$$

The following lemma is a special case of Lemma 3 in [29].

LEMMA 2.2. If $L: X \to X$ is given by (2.18), L is a compact linear operator. The spectrum of the complexification of L is given by

$$\sigma(L) = \{\lambda_p : p = \text{an odd, positive integer}, \quad \lambda_p = (-1)^{(\frac{p-1}{2})}(\tfrac{2}{p\pi})\} \cup \{0\} .$$

Each λ_p is a simple eigenvalue with corresponding eigenvector $x_p(t) = \sin((\frac{p\pi}{2})t)$.

We shall refer (in the obvious notation) to solutions (x,α) of (2.17), where $x \in X$ and $\alpha \in R$. If $f'(0) = 1$, Lemma 2.2 implies that the only possible places where bifurcation can occur are at

$$(2.19) \qquad \alpha = \alpha_p = (-1)^{(\frac{p-1}{2})}(\tfrac{p\pi}{2}) , \quad p = \text{an odd, positive integer}$$

(Recall that bifurcation occurs at $(0,\alpha)$ if there exists a sequence $(x_n,\beta_n) \to (0,\alpha)$ with $x_n \neq 0$ and $x_n = \beta_n F(x_n))$. On the other hand, since λ_p is a simple eigenvalue, standard results (see [10]) imply that bifurcation occurs at $(0,\alpha_p)$.

LEMMA 2.3. Assume that $f: R \to R$ is locally Lipschitzian and odd and that $f'(0) = 1$. Define $S \subset X \times R$ by

$$(2.20) \qquad S = \{(x,\alpha) \mid x = \alpha F(x) \text{ and } x \neq 0\}$$

and let $\mathcal{S}$ denote the closure of S in $X \times R$. Then we have

$$(2.21) \qquad \mathcal{S} = S \cup (\bigcup_{p\geq 1}(0,\alpha_p))$$

where the union in (2.21) is over odd, positive integers p . If $(x,\alpha) \in S$, define $z(x,\alpha)$ to be the number of real numbers t such that $0 < t < 2$ and $x(t) = 0$. Extend z to a map of $\mathcal{S}$ by defining $z(0,\alpha_p) = p-1$. Then $z: \mathcal{S} \to R$ is a continuous map.

PROOF. If $(x,\alpha) \in S$ and $y(t) = x(t-1)$, then $(x(t),y(t))$ satisfies (2.2). To show that z is continuous on S , it suffices to show that if $(x,\alpha) \in S$ and $x(t_0) = 0$, then $\dot{x}(t_0) \neq 0$. However, if $\dot{x}(t_0) = 0 = -\alpha f(y(t_0))$, uniqueness of solutions of the initial value problem for (2.2) would imply that $x(t) = x(t_0) = 0$ and $y(t) = y(t_0)$ for all t , a contradiction.

To prove continuity at $(0,\alpha_p)$, suppose that $(x_n,\beta_n) \in X \times R$ and (x_n,β_n) approaches $(0,\alpha_p)$. We will suppose that $z(x_n,\beta_n)$ does not approach $p-1$ and obtain a contradiction. By taking a subsequence (which we label the same) we can assume that

$$(2.22) \qquad |z(x_n,\beta_n) - (p-1)| \geq 1 .$$

In the notation of equation (2.16) we have

$$(2.23) \qquad x_n = \beta_n F(x_n) \ .$$

If we divide (2.23) by $\|x_n\|$, write $u_n = (\frac{x_n}{\|x_n\|})$ and recall that F has a Fréchet derivative L at 0, we find

$$(2.24) \qquad \begin{aligned} u_n &= \beta_n L(u_n) + \beta_n \frac{R(x_n)}{\|x_n\|} \ , \\ \lim_{n\to\infty} \frac{R(x_n)}{x_n} &= 0 \ . \end{aligned}$$

Since L is a compact linear operator, by taking a subsequence we can assume that $L(u_n)$ converges in X, so u_n converges to u in X, $\|u\| = 1$. Clearly, u must be an eigenvector of L with eigenvalue α_p^{-1}, and because α_p^{-1} is a simple eigenvalue we obtain

$$(2.25) \qquad u(t) = \pm \sin(\tfrac{p\pi}{2}t) \ .$$

Starting from the equation

$$\dot{x}_n(t) = -\alpha f(x_n(t-1))$$

one can also see that $\dot{u}_n(t)$ converges uniformly to $-\alpha_p u(t-1) = \dot{u}(t)$. Since $\dot{u}(t_o) \neq 0$ if $u(t_o) = 0$, it is now straightforward to show that u_n and u have the same number of zeros on $(0,2)$ for n large enough, namely $p-1$. This contradicts equation (2.22) and proves continuity of z at $(0,\alpha_p)$.

REMARK 2.2. If we assumed more differentiability of f, the continuity of z at $(0,\alpha_p)$ would follow from results of Crandall and Rabinowitz.

LEMMA 2.4. Let notation and assumptions be as in Lemma 2.3 and let C_p (p an odd, positive integer) denote the connected component of S which contains $(0,\alpha_p)$. Then C_p is unbounded in $X \times R$ and $C_p \cap C_q$ is empty for $p \neq q$. If there exists a number ζ_1 such that $f(\zeta_1) = 0$ and $f(x) > 0$ for $0 < x < \zeta_1$, then $\|x\| < \zeta_1$ for every $(x,\alpha) \in C_p$.

PROOF. Lemma 2.3 implies that

$$z(x,\alpha) = p-1 \ , \quad (x,\alpha) \in C_p$$

so we must have $C_p \cap C_q = \phi$ for $p \neq q$. In particular, $(0,\alpha_q) \notin C_p$, and Rabinowitz's global bifurcation theorem [37] now implies that C_p is unbounded.

If there exists $(x,\alpha) \in C_p$ with $\|x\| > \zeta_1$, the connectedness of C_p implies that there exists $(u,\beta) \in C_p$ with $\|u\| = \zeta_1$. Assume for the moment that $p = 1$. By replacing (u,β) by $(-u,\beta) \in C_1$ if necessary, we can assume (use lemma 2.3) that $u(t) > 0$ for $0 < t < 2$. It follows then from equation (2.1) that u achieves its maximum at $t = 1$, and equation (2.2) then implies that $u(t) = \zeta_1$ for all t, which contradicts $u(0) = 0$.

Next consider the case of general p, so $(u,\beta) \in C_p$ and $\|u\| = \zeta_1$. By replacing $(u,\beta) \in C_p$ by $(-u,\beta)$ if necessary, we can assume that $u(-1) < 0$. Lemma 2.1 implies that if

$$r = \inf\{t > 0 : u(t-1) = 0\} \tag{2.26}$$

then

$$\begin{cases} u(t+2r) = -u(t) & \text{for all } t \\ u(-t) = -u(t) & \text{for all } t \end{cases} \tag{2.27}$$

and either $u(t) > 0$ for $0 < t < 2r$ or $u(t) < 0$ for $0 < t < 2r$. We know that u has $(p-1)$ zeros on $(0,2)$ (p odd), and equation (2.27) implies that the separation between zeros is $2r$, so

$$p(2r) = 2 \quad \text{or} \quad r = \left(\frac{1}{p}\right) \tag{2.28}$$

Using (2.27), (2.28) and (2.1), one can see that u is increasing on $(-r,r)$, decreasing on $(r,3r)$, etc. It follows that $u(1) = \pm\zeta_1$, and the same proof as for the case $p = 1$ gives a contradiction.

<u>REMARK 2.3</u>. Using the above remarks it is not hard to see that C_p and C_1 are homeomorphic by the map $\varphi : C_p \to C_1$

$$\begin{cases} \varphi(x,\beta) = (\tilde{x},\tilde{\beta}) \\ \tilde{x}(t) = x(tp^{-1}), \quad \tilde{\beta} = (-1)^{\left(\frac{p-1}{2}\right)}\left(\frac{\beta}{p}\right) \end{cases} \tag{2.29}$$

<u>REMARK 2.4</u>. If K is as in (2.15), let $C_1^+ = \{(x,\alpha) \in C_1 : x \in K\}$ and $C_1^- = \{(x,\alpha) \in C_1 : -x \in K\}$. Our previous results show

$$\begin{cases} C_1 = C_1^+ \cup C_1^- \quad \text{and} \\ C_1^+ \cap C_1^- = \{(0,\frac{\pi}{2})\} \end{cases} \tag{2.30}$$

and one can easily derive from (2.30) that C_1^+ is connected.

Our next lemma provides <u>a priori</u> bounds on the positive special periodic solutions of (2.1). Deeper results in the same spirit will be obtained in section 4.

LEMMA 2.5. Let $g: \mathbb{R} \to \mathbb{R}$ be an odd, locally Lipschitzian map. Assume that

$$-M \le g(x) \le M_* \ , \quad 0 \le x \le R_o$$

and that there exist finite, positive numbers γ_1 and γ_2, $\gamma_1 < \gamma_2$, such that

$$\gamma_1 x \le g(x) \le \gamma_2 x \quad \text{for} \quad x \ge R_o$$

In addition suppose that $\gamma_1 > \frac{\pi}{2}$ or $\gamma_2 < \frac{\pi}{2}$. Then there exists a number R, which can be chosen to depend continuously on $(\gamma_1, \gamma_2, R_o, M, M_*)$ as above, such that if $x \in K$ satisfies $\dot{x}(t) = -g(x(t-1))$, then $\|x\| \le R$.

PROOF. As before, define $y(t) = x(t-1)$ and recall that $(x(t), y(t))$ satisfies

$$(2.31) \qquad \begin{cases} \dot{x}(t) = -g(y(t)) \\ \dot{y}(t) = g(x(t)) \end{cases}$$

and that if

$$G(u) := \int_0^u g(s)\,ds$$

$$(2.32) \qquad G(x(t)) + G(y(t)) = G(x(0))$$

We shall show that there exists a number R such that the assumption that $\|x\| \ge R$ and $x \in K$ leads to a contradiction. Thus assume that $\|x\| = R$, where R will be estimated later, and select t_o, $1 \le t_o < 2$, such that

$$x(t_o) = R .$$

Obvious estimates give

$$(2.34) \qquad \begin{cases} G(y(t_o)) \ge -MR_o \\ G(x(t_o)) \ge \gamma_1 \int_{R_o}^{R} s\,ds - MR_o \end{cases}$$

so we obtain from (2.32) that

$$G(x(t)) + G(y(t)) \ge \frac{\gamma_1}{2} R^2 - (\frac{\gamma_1}{2} R_o^2 + 2MR_o) := (\frac{\gamma_1}{2}) R^2 - B_o \tag{2.34}$$

Because $x(\frac{3}{2}) = y(\frac{3}{2})$, (2.34) implies that

$$2G(x(\tfrac{3}{2})) \ge (\frac{\gamma_1}{2})^2 R^2 - B_o \tag{2.35}$$

Our assumptions on g give

$$G(x) \le (\frac{\gamma_2}{2}) x^2 + M_* R_o$$

so we obtain from (2.35) that

$$(x(\tfrac{3}{2}))^2 \ge (\frac{\gamma_1}{2\gamma_2}) R^2 - B_1 \ , \ B_1 = \gamma_2^{-1}(B_o + 2M_* R_o) \tag{2.36}$$

Using (2.36), select R so large that $x(\frac{3}{2}) \ge 4R_o + M$. Define t_1 to be the first time $t \ge \frac{3}{2}$ such that $x(t) = R_o$. Equation (2.31) implies that $x(t)$ is decreasing and $y(t)$ increasing for $\frac{3}{2} \le t \le t_1$.

We want to estimate $2-t_1$. Crude estimates show that $\dot{y}(t) \ge -M$ for $1 \le t \le 2$, so

$$y(t) \ge y(\tfrac{3}{2}) - \tfrac{1}{2}M \quad \text{for} \quad \tfrac{3}{2} \le t \le 2 \tag{2.37}$$

Combining (2.36) and (2.37) and assuming that R is sufficiently large gives

$$\begin{aligned} & y(t) \ge \sqrt{(\frac{\gamma_1}{2\gamma_2}) R^2 - B_1} - \tfrac{1}{2}M := \rho \ge R_o \quad \text{for} \quad \tfrac{3}{2} \le t \le 2 \\ & \dot{x}(t) = - g(y(t) \le - \gamma_1 \rho \quad \text{for} \quad \tfrac{3}{2} \le t \le 2 \end{aligned} \tag{2.38}$$

Equation (2.38) implies that

$$\sigma_1 = 2 - t_1 \le \frac{R_o}{\gamma_1 \rho} \tag{2.39}$$

By symmetry about the line $y = x$ we find that there exists a number $\sigma \le \frac{R_o}{\gamma_1 \rho}$ such that

$$\begin{cases} x(t) \ge R_o & \text{for} \quad 1 \le t \le 2-\sigma_1 \\ y(t) \ge R_o & \text{for} \quad 1+\sigma \le t \le 2 \\ x(t) \le R_o & \text{for} \quad 2-\sigma_1 \le t \le 1 \\ y(t) \le R_o & \text{for} \quad 1 \le t \le 1+\sigma \end{cases} \tag{2.40}$$

Also we know that $y(t) \geq \rho$ on $[\frac{3}{2},2]$ and similarly for x on $[1,\frac{3}{2}]$.

(2.41)
$$\left\{\begin{aligned} &\text{If } \Theta(t) = \arctan\left(\frac{y(t)}{x(t)}\right) \text{ for } 1\leq t\leq 2 \text{, then we have} \\ &\frac{\pi}{2} = \int_1^2 \dot{\Theta}(t)dt = \int_1^2 [g(x)x+g(y)y](x^2+y^2)^{-1}dt \\ &= \int_1^{1+\sigma} \dot\Theta(t)dt + \int_{1+\sigma}^{2-\sigma_1} \dot{\Theta}(t)dt + \int_{2-\sigma_1}^{2} \dot\Theta(t)dt \\ &= I_1 + I_2 + I_3 \end{aligned}\right.$$

Assume for definiteness now that $\gamma_1 > \frac{\pi}{2}$; the proof in the case $\gamma_2 < \frac{\pi}{2}$ is analogous. Since $x(t) \geq R_o$ and $y(t) \geq R_o$ for $1+\sigma\leq t\leq 2-\sigma_1$ we find

(2.42)
$$I_2 \geq \int_{1+\sigma}^{2-\sigma_1} \frac{\gamma_1(x^2+y^2)}{x^2+y^2}\, dt = \gamma_1(1-\sigma-\sigma_1)$$

Similarly we find that (with ρ as in (2.38))

(2.43)
$$I_1 = \int_1^{1+\sigma} \left\{\frac{g(x)x+g(y)y}{x^2+y^2}\right\}dt \geq \int_1^{1+\sigma} \left\{\frac{-2MR_o}{x^2+y^2}\right\}dt \geq -\left(\frac{2MR_o}{\rho^2}\right)\sigma$$

and

(2.44)
$$I_3 \geq -\left(\frac{2MR_o}{\rho^2}\right)\sigma_1$$

Using (2.39), (2.42), (2.43) and (2.44) we obtain

(2.45)
$$\frac{\pi}{2} \geq \gamma_1 - (\sigma+\sigma_1)\left(\gamma_1+\frac{2MR_o}{\rho^2}\right) \geq \gamma_1 - \left(\frac{R_o}{\gamma_1\rho}\right)\left(\gamma_1+\frac{2MR_o}{\rho^2}\right)$$

Because $\gamma_1 > \frac{\pi}{2}$ and $\rho \to \infty$ as $R \to \infty$, (2.45) gives a contradiction for R large enough. Furthermore, an upper estimate on R can be obtained from (2.45) and can be seen to depend continuously on $\gamma_1 > \frac{\pi}{2}$, $\gamma_2 \geq \gamma_1$, M, M_* and R_o.

REMARK 2.5. In Lemma 2.5 we have eliminated the case $\gamma_2 = +\infty$. In fact, it is also possible to obtain *a priori* bounds if

$$\lim_{|y|\to\infty} \frac{g(y)}{y} = +\infty$$

and f is increasing on intervals $(-\infty,-R_o]$ and $[R_o,\infty)$. This is proved in [26] Lemma 2.5, under the added assumption that $xg(x)>0$ for $x \neq 0$ (but without assuming oddness). For reasons of length, we shall not pursue this point.

With the aid of the preceding lemmas we can obtain the main result of this section. We allow a nonlinearity f which cannot be linearized at ∞, however, we will show that our problem has bifurcation from ∞. The proof here will essentially be based on a global perturbation argument together with the Leray-Schauder Continuation Method. This idea is similar to an approach to obtain bifurcation from ∞ for nonlinear elliptic boundary value problems [36,34]. We first specify the nonlinearity:

(H) Let $f: R \to R$ be an odd, locally Lipschitzian function which is differentiable at 0 with $f'(0) = 1$. Assume that there exists $\eta_1 > 0$ such that $f(\eta_1) = 0$ and that there exists positive constants β_1 and β_2 and a number R_0 such that

$$\beta_1 \le f(x)x^{-1} \le \beta_2 \quad \text{for} \quad x \ge R_0$$

To define the global perturbation we let $g: R \to R$ be an odd, locally Lipschitzian function with

$$g(x) = \begin{cases} f(x) & , \text{ for } |x| \le (\frac{1}{2})\,\eta_1 \text{ and } x \ge R_0 \\ \text{linear function}, & \text{for } (\frac{1}{2})\,\eta_1 \le x \le R_0 \end{cases}$$

Now let $\alpha > 0$. Then the perturbation h_λ is just given by a one-parameter family which embeds f and g:

$$h_\lambda(x) = \begin{cases} \lambda f(x) & , \text{ for } \lambda \le \alpha \\ \alpha[(\alpha+1-\lambda)f(x)+(\lambda-\alpha)g(x)] & , \text{ for } \alpha \le \lambda \le \alpha+1 \\ (\lambda-1)g(x) & , \text{ for } \lambda \ge \alpha+1 \end{cases} \tag{2.46}$$

By assumption f is linearizable at 0 and, thus, (2.1) has a classical bifurcation from trivial solutions. The idea of the perturbation is to find a global continuum of periodic solutions for

$$\dot{x}(t) = -h_\lambda(x(t-1))$$

which yields bifurcation from ∞ for (2.1) (see figure 49).

THEOREM 2.1. Let f satisfy (H) . Then there exist continua Σ_o and Σ_∞ of positive, special periodic solutions of (2.1) satisfying:

(1) Σ_o bifurcates from 0 at $\lambda = \lambda_o = \pi/2$ and $(x,\lambda) \in \Sigma_o$ implies that $\|x\| < \eta_1$.

(2) Σ_∞ bifurcates from ∞ in $(0,\lambda_\infty^+]$ with $\lambda_\infty^+ = \pi/2\beta_1$, (i.e. there exists a sequence $(x_n,\lambda_n) \in \Sigma_\infty$ with $\|x_n\| \to \infty$ and $\lambda_n \in (0,\lambda_\infty^+ + n^{-1}])$ and $(x,\lambda) \in \Sigma_\infty$ implies that $\|x\| > \eta_1$.

(3) Let $\Pi: K \times R_+ \to R_+$ denote the projection onto the second component, then

$$\Pi(\Sigma_o) \supset (\lambda_o,\infty)$$
$$\Pi(\Sigma_\infty) \supset (\lambda_\infty^+ \infty)$$

(4) For any $\alpha > \max\{\pi/2\beta_1, \pi/2\}$ equation (2.1) has at least two positive, special periodic solutions in $K-\{0\}$ which lie in a connected set of solutions of

$$\dot{x}(t) = -h_\lambda(x(t-1)) .$$

PROOF. By selecting η_1 to be the first positive zero of $f(x)$, we can assume that $f(x) > 0$ for $0 < x < \eta_1$. Select $\alpha > \max\{\pi/2\beta_1, \pi/2\}$ and choose the perturbation h_λ according to (2.46). Define the operator $\Phi_\lambda : X \to X$ by

$$(2.47) \qquad (\Phi_\lambda x)(t) = -\int_0^t h_\lambda(x(s-1))ds$$

and define T by

$$T = \{(x,\lambda) \in X \times R : x = \Phi_\lambda(x), \quad x \neq 0\}$$

and take $\mathcal{T}$ to be the closure of T in $X \times R$.
If Σ denotes the connected component of $\mathcal{T}$ which contains $(0,\pi/2)$ and $\Sigma^+ = \Sigma \cap K \times R_+$, the same arguments as before imply that Σ^+ is unbounded and connected. To prove the theorem we first will prove the claims contained in figure 49. Observe that if $(x,\lambda) \in \Sigma^+$ and $x(t) = 0$ for some t with $0 < t < 2$, then $x \equiv 0$ and $\lambda = \pi/2$.

We next claim that there exists $\lambda_1 > \alpha + 1$ such that if $\lambda \geq \lambda_1$ then $\Phi_\lambda(x) = x$ has no nontrivial solutions in K . To see this, introduce a partial ordering on K by $u \leq v$ if $u(t) \leq v(t)$ for $0 \leq t \leq 2$. Select $\delta > 0$ so that $\delta x \leq g(x)$ for all $x \geq 0$ and note that for all

$u \in K$ and $\lambda \geq \alpha + 1$.

$$\Phi_\lambda u \geq A_\lambda u := (\lambda-1)\delta Lu \tag{2.48}$$

where L is defined by (2.18). We now use a standard argument from the theory of positive operators. Select $\lambda_1 \geq \alpha + 1$ such that the spectral radius of A_{λ_1} is a number $r > 1$. The linear Krein-Rutman theorem implies that there exists $v \in K - \{0\}$ such that

$$A_{\lambda_1} v = rv \tag{2.49}$$

If $\Phi_\lambda(u) = u$ for some $u \in K - \{0\}$, $\lambda \geq \lambda_1$, it is not hard to see that there exists $\varepsilon > 0$ such that

$$u \geq \varepsilon v \tag{2.50}$$

It follows from (2.50) that

$$u = \Phi_\lambda(u) \geq A_\lambda u \geq \varepsilon A_\lambda v$$

and in general (use induction) that

$$u \geq \varepsilon A_\lambda^n v = \varepsilon r^n v \tag{2.51}$$

Since $r > 1$, equation (2.51) gives a contradiction.

Let γ be any fixed constant such that

$$\gamma > \frac{\pi}{2\beta_1} \quad \text{and} \quad \gamma < \alpha$$

Lemma 2.5 implies that any solution of $x = \Phi_\lambda(x)$ with $\lambda \geq \gamma$ (so $\lambda \leq \lambda_1$) must satisfy

$$\|x\| \leq R$$

where R depends on γ . The same argument used in the proof of lemma 2.4 also shows that for all $(x,\lambda) \in \Sigma^+$ with $\lambda \leq \alpha$ we have that $x \neq \eta_1$. Thus, we have shown that there exists a continuum Σ^+ as given in figure 49 and this yields (4) by a simple connectedness argument. Moreover, the construction of Σ^+ implies that problem (2.1) has a bifurcation from ∞ for positive, special periodic solutions in $(0,\lambda_\infty^+]$. It remains to show the existence of Σ_0 and Σ_∞ . The existence of Σ_0 is obvious from Lemma 2.4 and the fact that λ_0 is a point of bifurcation from zero. The existence of Σ_∞ follows essentially because in our perturbation the choice of γ and α were arbitrary, except that

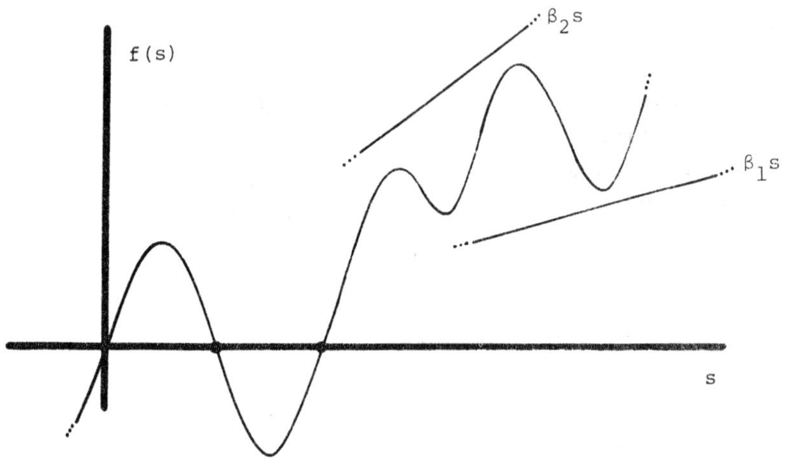

Figure 48. (nonlinearity)

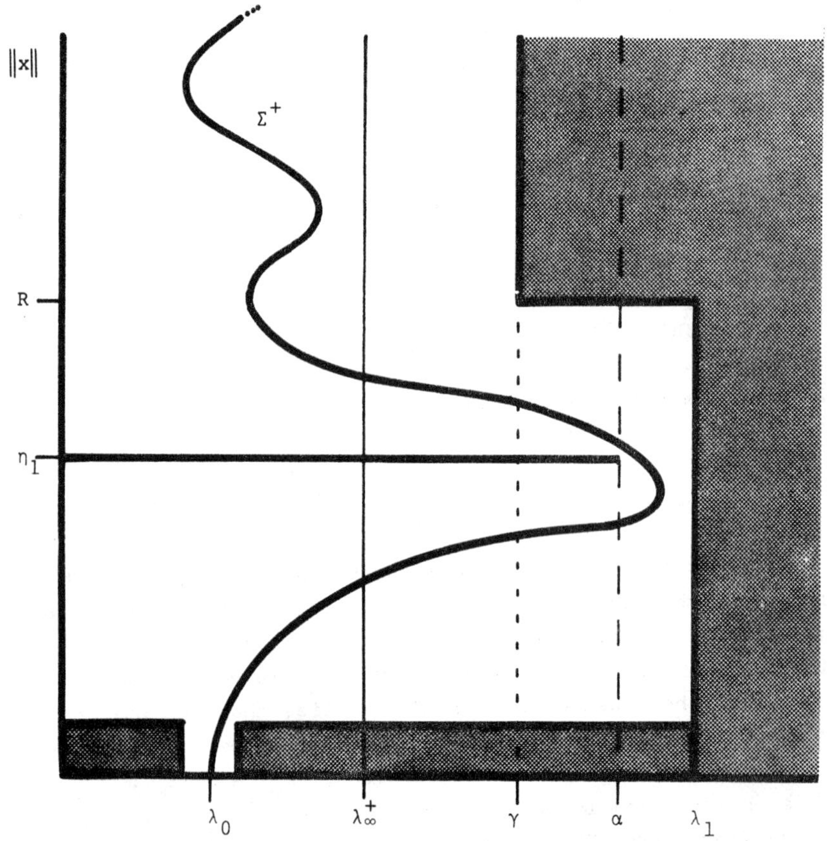

Figure 49. (bifurcation diagram for perturbation)

$$\alpha > \max\{\pi/2\beta_1, \pi/2\} \quad \text{and} \quad \pi/2\beta_1 < \gamma < \alpha ;$$

thus, one can take a sequence $\alpha_n \geq n$. Then one uses arguments which are more or less standard to obtain Σ_∞ (cf.[40]).

REMARK 2.6. Using the *a priori* estimates on positive, special periodic solutions and the perturbation idea of Theorem 2.1 it is now easy to obtain also results for nonlinearities f which satisfy (H) except that $f'(0) = -1$, see figure 50.

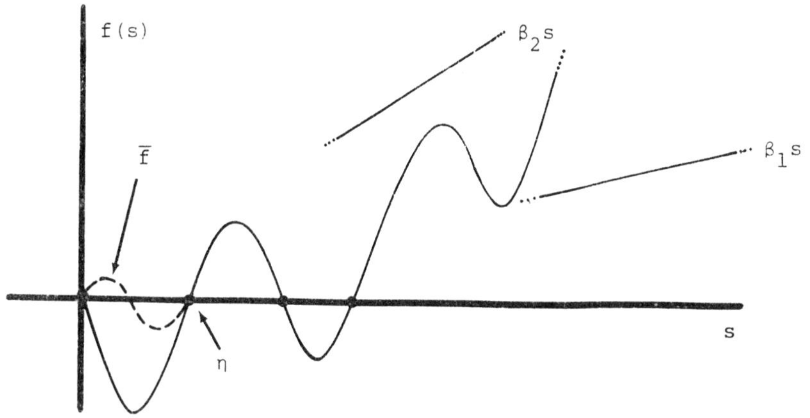

Figure 50.

Here we do not have bifurcation from zero for positive, special periodic solutions. But we still can obtain a continuum Σ_∞ of positive, special periodic solutions of (2.1), where f is as in figure 50. To see this we just extend the perturbation h_λ in the following way. We choose $\alpha > \pi/2\beta_1$ and choose a function $\bar{f}$ which satisfies (H) and which agrees with f for all $x \geq \eta$, where η is the first positive zero of f. Now choose a one-parameter family h_λ with

$$\begin{aligned} h_\alpha &= f \\ h_{\alpha+1} &= \bar{f} \\ h_{\alpha+2} &= g \end{aligned}$$

and obtain a continuum Σ^+ for the perturbed equation as before. Letting $\alpha \to \infty$ we obtain Σ_∞.

CHAPTER 3

3. NUMERICAL APPROXIMATION AND SPURIOUS SOLUTIONS

The problem of finding numerical solutions which are approximations of the special periodic solutions of

$$\dot{x}(t) = -\alpha f(x(t-1)) \tag{3.1}$$

can be attacked in many different ways. For example, at first glance it seems to be reasonable to make use of the cyclic system

$$\begin{cases} \dot{x}(t) = -\alpha f(y(t) \\ \dot{y}(t) = \alpha f(x(t)) \end{cases} \tag{3.2}$$

and use an higher order initial value problem solver to compute periodic solutions of (3.2). As we have seen in chapter 1 typical implicit one-step methods which inherit many of the structural properties intrinsic to (3.2) may fail because of the typical appearance of a transversal homoclinic structure. Another reasonable approach seems to be to exploit the integral equation (2.16) from chapter 2 for a numerical scheme:

$$\begin{cases} (Fx)(t) := -\int_0^t f(x(s-1))ds \\ x = \alpha F(x) \end{cases} \tag{3.3}$$

Recall that $F: X \to X$ is a continuous, compact map and

$$x = \alpha F(x)$$

implies that x is a special periodic solution of (3.1) when defining the Banach space X to be

$$X = \{x : \mathbb{R} \to \mathbb{R} \ : \ x(t+2) = -x(t) \ , \ x(-t) = -x(t) \text{ and } x \text{ continuous}\} \ ,$$

equipped with the norm $||x|| = \sup |x(t)|$. We will discuss in the following a discrete approximation of (3.3) which is based on the composite trapezoidal rule and we will show using in an essential way the phase plane studies of chapter 1 for (3.2) that there are several types of numerical solutions which do not approximate a solution of (3.1). These <u>spurious</u> solutions will be classified and at the end of this chapter numerical experiments will be discussed. The choice of the trapezoidal rule which is a second order method is only for convenience and the avoidance of technicalities.

First we derive the numerical approximation of (3.3). This is achieved by using the composite trapezoidal rule to approximate the integral operator and then use the symmetries of X and oddness of f to obtain a system of n equations from $x = \alpha F(x)$ (see figure 51)

$$(3.4) \qquad Ax - \mu B\Phi(x) = 0 \ , \quad x \in R^n \ , \quad \mu \in R$$

where $h = {}^1/n$, $t_i = i \cdot h$ $(i = 0, 1, \ldots, n)$, $x_i = x(t_i)$, $\mu = \alpha h/_2$,

$$A = \begin{pmatrix} 1 & & & \bigcirc \\ -1 & \ddots & & \\ & \ddots & \ddots & \\ \bigcirc & & -1 & 1 \end{pmatrix} , \qquad B = \begin{pmatrix} \bigcirc & & 1 & 1 \\ & \cdot^{\cdot^{\cdot}} & \cdot^{\cdot^{\cdot}} & \\ 1 & \cdot^{\cdot^{\cdot}} & & \\ 1 & & & \bigcirc \end{pmatrix}$$

and $\Phi(x) = (f(x_1), \ldots, f(x_n))^T$.

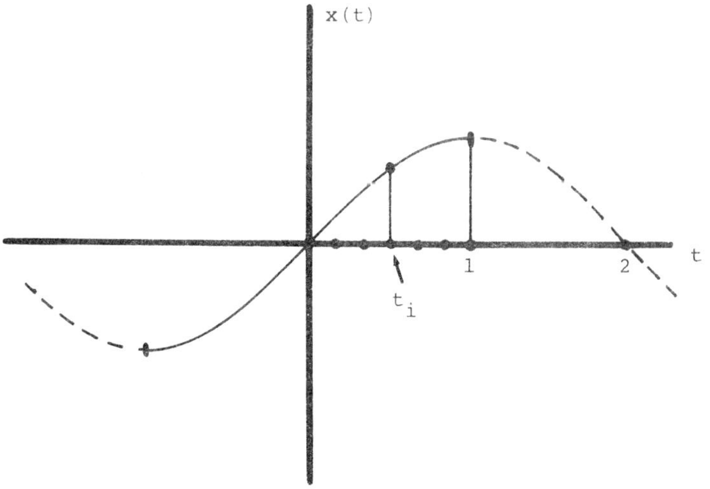

Figure 51.

For convenience we will assume throughout this chapter that f satisfies (1.2) and is C^1 . We are only interested in periodic solutions of (3.1) which lie in X and so that $x(t) > 0$ for $0 < t < 2$. Thus, we restrict (3.4) to the open cone $\overset{\circ}{K}$, where

$$K = R_+^n = \{x \in R^n : x_i \geq 0 \ , \quad i = 1, \ldots, n\} .$$

Observe that B^{-1} exists and that if $B^{-1} = (\beta_{ij})$ then

$$\beta_{ij} = \begin{cases} 0 & , \ i+j \le n \\ (-1)^{i+j-n-1} & , \ i+j > n \end{cases}$$

Thus, (3.4) is equivalent to

$$(3.5) \qquad \begin{cases} G(x,\mu) = 0 \ , \quad x \in \overset{\circ}{K} \ , \quad \mu \in R \\ \text{with} \\ G(x,\mu) = \mu^{-1}B^{-1}Ax - \Phi(x) \end{cases}$$

Observe that the system $G(x,\mu) = 0$ has a nonlinear part $\Phi(x)$ which is decoupled, i.e.

$$\Phi(x) = 0 \Leftrightarrow x_i \in f^{-1}(0) \ ,$$

and the linear operator plays the role of a coupling matrix. To solve (3.5) for $\mu >> 1$ is therefore a simple application of the implicit function theorem:

LEMMA 3.1. Let k be the number of positive simple zeros of f. Then there exists $M_1 > 0$, such that for all $\mu \ge M_1$ problem (3.5) has at least k^n solutions in $\overset{\circ}{K}$. Furthermore, these solutions are arranged on smooth curves $C_i \subset \overset{\circ}{K}$, $i = 1, \ldots, k^n$, which are parametrized over $[M_1,\infty)$ and for which

$$C_i(\mu) \to z^i \quad \text{as} \quad \mu \to \infty$$

$$\text{and} \quad \Phi(z^i) = 0 \ .$$

In view of a comparison with the results in chapter 2 and 3 (see especially remark 1.2) we assume additionally that

$$(3.6) \qquad \begin{cases} f(s) = m_\infty s + \phi_\infty(s) \ , \ \text{with} \ \ \phi_\infty(s)/s \to 0 \ \text{as} \ s \to \infty \\ f(s) = m_o s + \phi_o(s) \ , \ \text{with} \ \ \phi_o(s)/s \to 0 \ \text{as} \ s \to 0 \\ m_o > 0 \ , \ m_\infty > 0 \ . \end{cases}$$

We also observe that A^{-1} exists and that if $A^{-1} = (\alpha_{ij})$ then

$$\alpha_{ij} = \begin{cases} 0 & , \quad i < j \\ \\ 1 & , \quad i \geq j \end{cases} .$$

Thus, $Ax - \mu B\Phi(x) = 0 \iff x = \mu A^{-1} B\Phi(x)$, where $A^{-1}B$ is the matrix

$$A^{-1}B = \begin{pmatrix} \bigcirc & & & 11 \\ & & \cdots & 21 \\ & \cdots & 2 & \vdots \\ 1 & \cdots & & \vdots \\ 2 & \cdots \cdots \cdots & 21 \end{pmatrix} ,$$

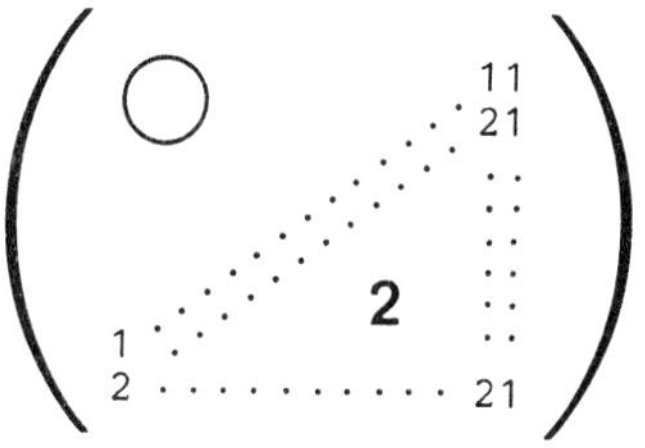

i.e. $A^{-1}B(K) \subset K$. According to the Perron-Frobenius theorem we have a positive eigenvalue λ^* of $D = A^{-1}B$ corresponding to a positive eigenvector $x^* \in K$. From the special structure of D it then easily follows that in fact $x^* \in \overset{o}{K}$. Using this fact one could straight forwardly prove the assertion of the following lemma but we prefer here to simply give a reference to [21] .

LEMMA 3.2. The eigenvalue λ^* of D is a simple eigenvalue of D and is the only eigenvalue corresponding to an eigenvector in K .

PROOF. One shows that with $u_0 = (1,\ldots,1) \in \overset{o}{K}$ for every $x \in K$ there are $\alpha, \beta > 0$ such that

$$\alpha u_0 \leq D^2 x \leq \beta u_0$$

($\leq$ the partial order induced by K). Thus, D is u_0-positive in the sense of [21] and the simplicity of λ^* follows from Theorem 2.10 in [21]. The uniqueness follows from Theorem 2.11 in [21] .

In view of a comparizon with the results in chapter 2 and 3, especially remark 1.2, we now have

THEOREM 3.1. Assume that f satisfies (1.2) and (3.6).

(1) There exists a critical parameter

$$\mu_o = (\lambda^* m_o)^{-1}$$

such that (3.5) has a continuum $S_o \subset R^{n+1}$ of solutions bifurcating at $\mu = \mu_o$ from the trivial solutions, and μ_o is the only possible point of bifurcation from trivial solutions in $K \times R$.

(2) There exists a critical parameter

$$\mu_\infty = (\lambda^* m_\infty)^{-1}$$

such that (3.5) has a continuum $S_\infty \subset R^{n+1}$ of solutions bifurcating at $\mu = \mu_\infty$ from infinity, and μ_∞ is the only possible point of bifurcation from infinity.

(3) For any $\varepsilon > 0$ there exists $R = R(\varepsilon) > 0$ such that for all solutions $(x,\mu) \in K \times R_+$ with

$$|\mu_\infty - \mu| \geq \varepsilon$$

one has that

$$\|x\| \leq R .$$

(4) Let $(x(\mu),\mu) \in K \times R_+$ be a family of solutions of (3.5) with $\mu \to \infty$. Then for any $\varepsilon > 0$ there exists $M_3 > 0$ such that for all $\mu \geq M_3$

$$\|x(\mu) - z\| < \varepsilon$$

for some $z \in \overset{o}{K}$ with $\Phi(z) = 0$.

PROOF. Set $H(x,\mu) = x-\mu A^{-1}B\Phi(x)$. Then $H_x(0,\mu) = Id-\mu m_o A^{-1}B$ and $H_x(\infty,\mu) = Id-\mu m_\infty A^{-1}B$. Thus, the bifurcation results in (1) and (2) follow, since λ^* is simple, from standard bifurcation theory developed by P. Rabinowitz [37]. To see that e.g. μ_o is the only possible point of bifurcation from trivial solutions in $K \times R$, assume that $(x^k, \mu^k) \in K \times R_+$ is a sequence of solutions of (3.5) with $x^k \to 0$ and $\mu^k \to \bar{\mu}$ and deduce from that an eigenvector

$$\bar{x} = \lim_{k\to\infty} x^k / \|x^k\| \in K$$

of $A^{-1}B$ corresponding to the eigenvalue $(\bar{\mu} m_0)^{-1}$ and then use the uniqueness of λ^*. The same type of argument holds for μ_∞. To prove (3) assume the contrary. Using (2) this means that we find a sequence $(x^k, \mu^k) \in K \times R_+$ with $x^k \to \infty$ and $\mu^k \to \infty$. Then we obtain a contradiction because

$$A \frac{x^k}{\|x^k\|} = \frac{\mu^k B\Phi(x^k)}{\|x^k\|} ,$$

and the left side is bounded while the right side is an unbounded sequence (because $m_\infty > 0$). Finally, assertion (4) follows because according to (3) $\|x(\mu)\| \leq R$ and therefore,

$$\Phi(x(\mu)) = {}^1/_\mu \, B^{-1}Ax(\mu) ,$$

where the right side goes to zero as $\mu \to \infty$.

Typically, one will have figure 52 assuming case (1.9) or (1.10) as in chapter 1.

We distinguish at least two types of spurious solutions. To compare the numerical solutions $x = (x_1, \ldots, x_n)$ of (3.5) with continuous time solutions of (3.2), we let $(x(t), y(t))$ be a solution of (3.2) for some $\alpha > 0$. Then according to our analysis in chapter 1 this solution is uniquely determined by the phase flow Φ^t subject to the initial value $(x(0), 0)$ on the x-axis. If we assume that we deal with case (1.9) or case (1.10) then we have that

$$\begin{cases} \text{either} & 0 < x(0) < z_1 \\ \text{or} & x(0) > r_2 . \end{cases}$$

Using the relation $x(t) = y(t-1)$ for this solution of period 4 we observe that

$$x(0) \text{ corresponds to } x_n ,$$

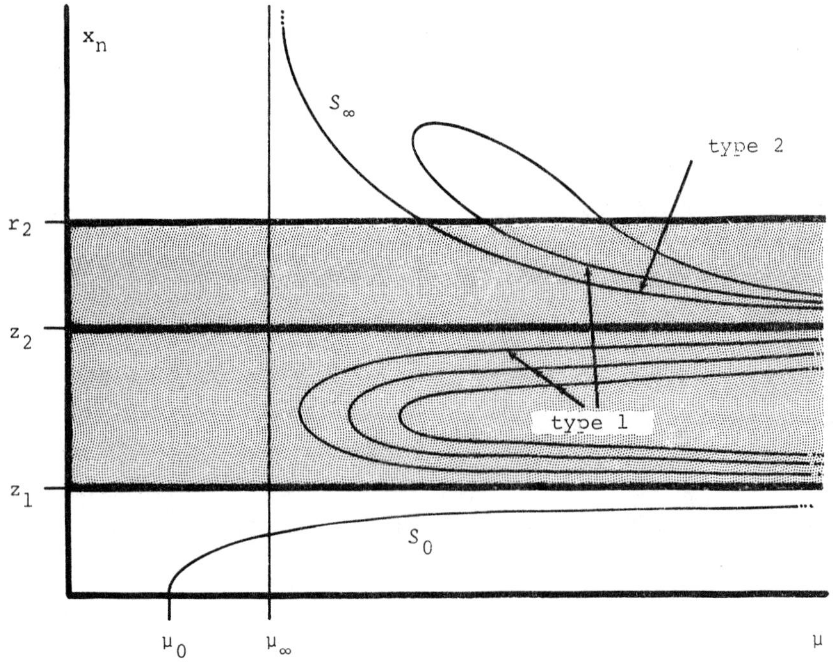

Figure 52. (spurious solutions)

where x_n is the last component in the numerical solution of (3.5). Thus, we have the following types

Type 1: These spurious solutions have an x_n component with $z_1 < x_n < r_2$ for μ sufficiently large and the corresponding continua are bounded in x_n for all μ.

Type 2: These spurious solutions have an x_n component with $z_2 < x_n < r_2$ for μ sufficiently large and the corresponding continuum is unbounded in x_n as $\mu \to \mu_\infty$.

This distinction is justified by the following argument. The phaseplane studies in chapter 1 suggest that under suitable assumptions on f equation (3.2) will have precisely two solutions of period 4, one will $x(0) < z_1$ and one with $x(0) > r_2$. In contrast (3.5) has at least k^n solutions in $\overset{\circ}{K}$. In our example for f (case (1.9) or case (1.10)) this means that we have

$$2^n - 1$$

spurious solutions, for $\mu >> 1$, i.e. all solutions with $x_n > z_1$ are spurious, i.e. the branch S_∞ which is guaranteed for (3.2) by theorem (3.1) will become spurious as $\mu \to \infty$. Thus, if we compare our numerical approaches in chapter 1 and here we see that in both cases

- approximation based on the cyclic system (3.2)

- approximation based on the integral equation (3.3)

one will have spurious solutions. So far, the only way to judge a given numerical solution is by understanding the master equation (3.1) sufficiently well and this will be increasingly difficult as f has a large number of zeros. Finally, we present some numerical experiments for (3.5). We discuss two nonlinearities

$$f_1(s) = \begin{cases} {}^1/_\pi \sin(\pi s) & , \ 0 \le s \le 2 \\ s - 2 & , \ s \ge 2 \end{cases}$$

$$f_1(-s) = - f_1(s)$$

$$f_2(s) = {}^s/_3(2+\cos(\tfrac{\pi}{2}\, 5))(s^2-1)(s^2-9)(s^4+9)^{-1}$$

Figure 53 shows f_2. Note that $f_1'(\infty) = 1 = f_1'(0)$; while $f_2'(\infty)$ does not exist. However, f_2 satisfies hypothesis (H) with $\beta_1 = {}^1/_3$ and $\beta_2 = {}^3/_3$.

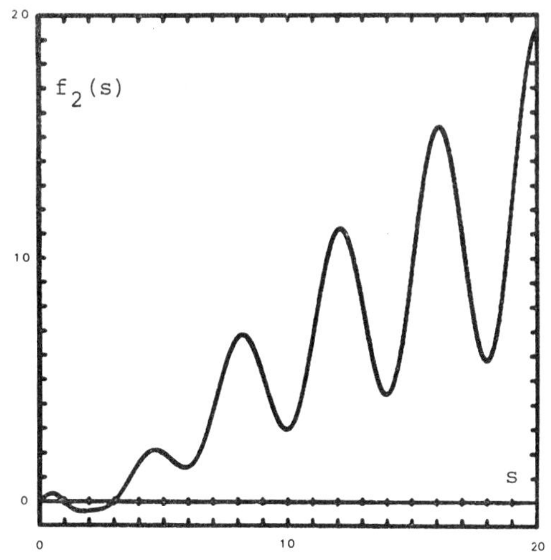

Figure 53. (nonlinearity f_2)

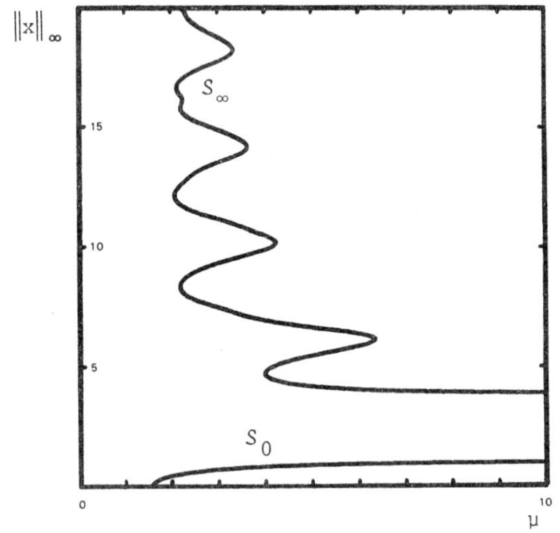

Figure 54.

Figure 54 shows a bifurcation diagram of (3.5) for the continua S_o and S_∞ (see theorem 3.1) with nonlinearity f_2. The norm is $\|x\|_\infty = \max |x_i|$ and $n = 30$.

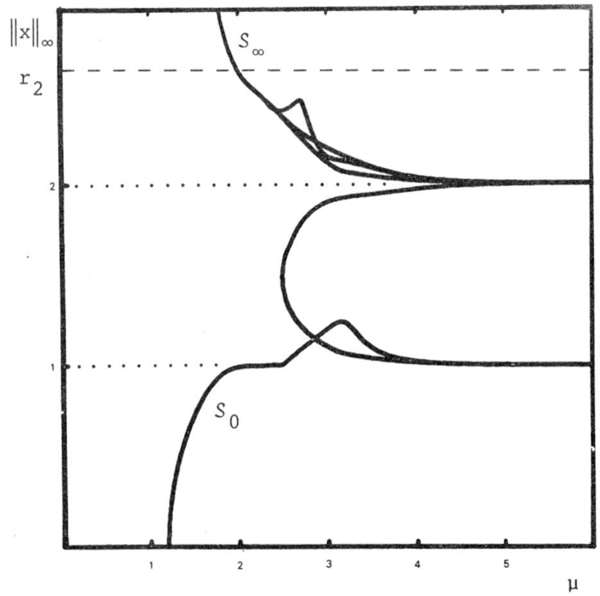

Figure 55.

Figures 55, 56 show the continua S_0 and S_∞ together with two continua of type 1 . In figure 55 we have used the norm $\|x\|_\infty = \max |x_i|$, while figure 56 shows the same continua with the norm $\|x\|_1 = \frac{1}{n}\sum_{k=1}^{n} x_k$. In both figures the nonlinearity is f_1 .

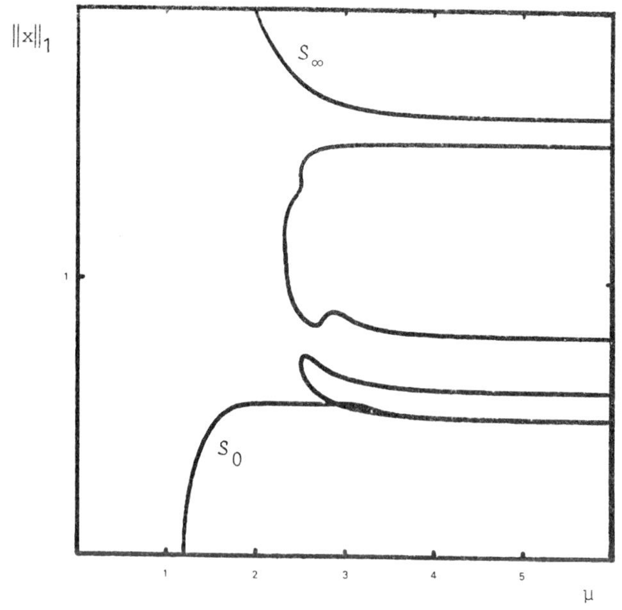

Figure 56.

Figure 55 demonstrates that indeed $x_n(\mu) \to z_i$ as $\mu \to \infty$ for all continua, where $z_1 = 1$ and $z_2 = 2$. Note that f_1 corresponds to case (1.11) in chapter 1 and one computes that

$$r_2 = \frac{2\pi+2}{\pi} \approx 2.64 \quad .$$

With the choice of $n = 10$ S_∞ drops below the level r_2 already for $\mu \geq 2$ and becomes spurious. It also seems to be remarkable that S_0 becomes spurious for $\mu \geq 2.5$ by exceeding the level $z_1 = 1$. Similar results for boundary value problems and finite difference approximations have been obtained in [33,35].

CHAPTER 4

4. A PRIORI BOUNDS WITHOUT THE ASSUMPTION OF ODDNESS

The proofs of our previous theorems all depended strongly on the oddness of the function $g(x)$ in the equation

$$(4.1) \qquad \dot{x}(t) = -g(x(t-1))$$

In this section we shall no longer assume that g is odd (although our previous assumptions on g will be strengthened in other directions), and we shall obtain *a priori* bounds on slowly oscillating periodic solutions of (4.1). One corollary of our theorems is the existence of periodic solutions of (4.1) for a new class of functions g. In fact Theorem 4.2 provides a partial answer to a conjecture raised in [26, Remark 2.2].

The details of our arguments will be rather technical, so we shall first describe the central idea. Phase plane ideas have previously been used to study (4.1): see [20,30,44]. On the other hand various tools from nonlinear functional analysis (especially the fixed point index for maps of a cone to itself) have also been used to study functional differential equations [27,28]. Here we combine these approaches for the first time. Phase plane methods yield the essential *a priori* bounds (Theorem 4.1), and a fixed point index argument is then used to obtain Theorem 4.2. Theorem 4.2 appears inaccessible with either method separately.

All results in this section will concern "slowly oscillating" periodic solutions of (4.1). Recall that a periodic solution $x(t)$ of (4.1) is called "slowly oscillating" if there exist numbers z_1 and z_2 with $z_1 > 1$ and $z_2 - z_1 > 1$ such that $x(0) = 0$, $x(t) > 0$ for $0 < t < z_1$, $x(t) < 0$ for $z_1 < t < z_2$ and $x(t+z_2) = x(t)$ for all t. Our first lemma is a weak geometrical statement about the curve $(x(t), \dot{x}(t))$ in R^2 when $x(t)$ is a slowly oscillating periodic solution of (4.1) of large norm.

LEMMA 4.1. (The inner disc lemma). Assume that $g: R \to R$ is a continuous map such that

$$(4.2) \qquad \lim_{x\to+\infty} g(x) = +\infty = \lim_{x\to-\infty} -g(x)$$

For every sufficiently large positive number M there exists a positive number $R = R(M,g)$ such that

$$(4.3) \qquad \lim_{M\to+\infty} R(M,g) = +\infty$$

and such that if $x(t)$ is a slowly oscillating periodic solution of (4.1) with

$$(4.4) \qquad \|x\| \equiv \sup_t |x(t)| \geq M$$

then for all t

$$(x(t))^2 + (\dot{x}(t))^2 \geq R^2 \tag{4.5}$$

The number R can be estimated as follows: Let A be a number such that $xg(x) > 0$ for $|x| \geq A$ and let B_1 and B_2 be such that

$$\begin{aligned} -B_1 \leq g(x) \leq B_2 \quad &\text{for } 0 \leq x \leq A \\ -B_2 \leq g(x) \leq B_1 \quad &\text{for } -A \leq x \leq 0 . \end{aligned} \tag{4.6}$$

Define functions h, γ and ψ by

$$\begin{cases} h(u) \equiv \max\{|g(x)| : |x| \leq u\} \\ \gamma(u) \equiv \inf\{|g(x)| : |x| \geq u\} \\ \psi(w) \equiv \inf\{u \geq 0 : h(u) = w\} . \end{cases} \tag{4.7}$$

If M is as in (4.4), define $M' = \psi(M-B_1)$ (assuming $M \geq B_1$), and assume M is large enough that the following inequalities have a solution M_*:

$$M \geq M_* \quad \text{and} \quad M' \geq M_* \tag{4.8a}$$

$$M \geq B_1 + M_* + h(M_*) \quad \text{and} \quad M' \geq B_1 + M_* + h(M_*) \tag{4.8b}$$

$$M_* > B_1 + A \tag{4.8c}$$

Then any slowly oscillating periodic solution $x(t)$ of (4.1) with $x \geq M$ satisfies (for all t)

$$(x(t))^2 + (\dot{x}(t))^2 \geq \min\left((\gamma(M_*-B_1))^2 , (M_*-B_1)^2\right) \tag{4.9}$$

<u>PROOF</u>. Let $x(t)$ be a slowly oscillating periodic solution of (4.1) with $\|x\| \geq M$ and suppose that z_1 and z_2 are the first and second zeros of $x(t)$ (as in the definition of slowly oscillating). Assume for definiteness that

$$M = \sup_{0 < s < z_1} x(t)$$

and that $x(s_1) = M$, $0 < s_1 < z_1$; the proof when $|x(t)|$ achieves its maximum on $[z_1, z_2]$ is analogous. A simple limiting argument shows that we can assume that the inequalities (4.8a) - (4.8c) are all strict.

We first claim that $s_1 < 2$ and
$$x(1) \geq M - B_1$$

To see this recall that $\dot{x}(t) \le B_1$ for $1 \le t \le z_1 + 1$ and $\dot{x}(t) \ge -B_1$ on $[z_1 + 1, z_2 + 1]$. If $s_1 > 2$, we would have

$$(4.10) \qquad x(t) \ge M - B_1 > A$$

for $s_1 - 1 \le t \le s_1$. Inequality (4.10) would imply that

$$(4.11) \qquad \dot{x}(s_1) = -g(x(s_1 - 1)) < 0$$

which contradicts the fact that x has a local maximum at s_1 . Since $s_1 < 2$, inequality (4.10) implies that

$$(4.12) \qquad x(1) \ge M - B_1$$

If $-M_2 \equiv \min_{z_1<t<z_2} x(t)$, one has

$$(4.13) \qquad M - B_1 \le x(1) = x(z_2 - 1) = \int_{z_2-1}^{z_2} |g(x(s))|ds \le h(M_2)$$

and (4.13) implies that

$$(4.14) \qquad M_2 \ge M' \equiv \psi(M - B_1)$$

Since we assume that $M' - B_1 > A$, an argument like that in the preceding paragraph proves

$$(4.15) \qquad |x(z_1 + 1)| \ge M' - B_1$$

We shall now prove inequality (4.9) for $1 \le t \le z_1 + 1$; the proof for the case $z_1 + 1 \le t \le z_2 + 1$ is analogous, and we shall omit it. Because $x(t)$ has period z_2 , it suffices to establish (4.9) for $t \in [1, z_2 + 1]$. Define numbers r_1 and t_1 by

$$r_1 = \inf\{t > 0 \mid x(t) = M_*\}$$

$$t_1 = \inf\{t > s_1 \mid x(t) = M_*\}$$

Straightforward estimates show that

$$(4.16) \qquad \begin{cases} x(t) \ge M_* - B_1 & \text{for } r_1 \le t \le 1 \\ x(t) \ge x(1) - h(M_*) & \\ \quad \ge M - B_1 - h(M_*) & \text{for } 1 \le t \le r \ + 1 \end{cases}$$

By using (4.16) and (4.8b) (strict form of the inequality) we see that

$$x(t) > M_* \quad \text{for} \quad 1 \le t \le r_1 + 1$$

and $t > r_1 + 1$.

Because we assume that $M_* - B_1 > A$ the above estimates show that

$$\dot{x}(t) = -g(x(t-1)) \le -\gamma(M_* - B_1) < 0 \quad \text{for} \quad r_1 + 1 \le t \le t_1+1. \tag{4.17}$$

It follows that if we define

$$R = \min\,(\gamma(M_* - B_1)\ ,\ M_* - B_1)$$

we have

$$(x(t))^2 + (\dot{x}(t))^2 \ge R^2 \quad \text{for} \quad 1 \le t \le t_1 + 1 .$$

Suppose we can prove that $x(t_1 + 1) \le -M_*$. Then an obvious estimate shows that

$$x(t) \le -M_* + B_1 \qquad \text{for} \quad t_1 + 1 \le t \le z_1 + 1$$

and we are done.

Thus it suffices to prove that $x(t_1 + 1) \le -M_*$. We assume not and obtain a contradiction. If $x(t_1 + 1) > -M$, we have $|x(t)| \le M_*$ for $t_1 \le t \le t_1 + 1$ (because $x(t)$ is decreasing on $[t_1, t_1 + 1]$) . We consider two cases.

<u>Case 1</u>. $-M_* < x(t_1 + 1) \le 0$. In this case we have $t_1 < z_1 \le t_1 + 1$ and for $t_1 + 1 \le t \le t_1 + 2$ we obtain

$$\begin{aligned} x(t) &= x(t_1 + 1) - \int_{t_1}^{t-1} g(x(s))ds \\ &\ge -M_* - h(M_*) \end{aligned} \tag{4.18}$$

and (4.18) implies that

$$|x(z_1 + 1)| \le M_* + h(M_*) . \tag{4.19}$$

Inequalities (4.15) and (4.19) imply

$$M' - B_1 \le M_* + h(M_*)$$

which contradicts (4.8b).

<u>Case 2</u>. $0 < x(t_1 + 1) \le M_*$. Define τ to be the first time $t \ge t_1$ such that $x(t) = A$. We know that $\dot{x}(t) < 0$ for $t_1 \le t \le t_1+1$,

and it is clear that $\dot{x}(t) < 0$ for $t_1 + 1 \le t \le \tau + 1$. We claim that

$$x(\sigma) \equiv \sup_{\tau \le t \le z_1} x(t) \le A + B_1$$

If this were not the case, we would have $x(\sigma) > A + B_1$ (where $\sigma > \tau+1$), and the usual argument would give

$$x(\sigma - 1) \ge x(\sigma) - B_1 > A$$

We would then find that

$$\dot{x}(\sigma) = - g(x(\sigma - 1)) < 0$$

Contradicting the fact that $\dot{x}(\sigma) = 0$.

We conclude from the above reasoning that

$$0 \le x(t) \le \max (M_*, A + B_1) = M_* \quad \text{for } t_1 \le t \le z_1 .$$

It follows that

$$M' - B_1 \le x(z_1 + 1) = \int_{z_1-1}^{z_1} g(x(s))ds \le h(M_*)$$

which contradicts (4.8).

Thus we have shown that $x(t_1 + 1) \le - M_*$, which completes the proof of the lemma.

In some cases it is possible to estimate the number R in Lemma 4.1 more explicitly.

LEMMA 4.2. Let g, B_1, B_2 and A be as defined in Lemma 4.1. Assume that there exist positive constants k_1 and k_2 such that

$$(4.20) \qquad k_1|x| \le |g(x)| \le k_2|x| \quad \text{for } |x| \ge A$$

and define $k_3 = \max (k_2,1)$. Let M be any positve number such that

$$(4.21) \qquad M > (k_2 + 1)k_3 (B_1 + A) + (k_3 + 1)B_1$$

and $\qquad M > B_1 + \max (B_1,B_2)$

and define M_* by

$$(4.22) \qquad M_* \equiv (k_2 + 1)^{-1}k_3 [M - (k_3 + 1)B_1]$$

If $x(t)$ is any slowly oscillating periodic solution of (4.1) such that

$$\sup_t |x(t)| \geq M \tag{4.23}$$

then one has

$$\inf_t [(x(t))^2 + (\dot{x}(t))^2] \geq [\min(k_1^2, 1)](M_* - B_1)^2 \tag{4.24}$$

PROOF. If h, γ and ψ are defined as in (4.7), it is easy to prove (using (4.20)) that

$$\begin{cases} h(u) \leq k_2 u & \text{for } u \geq A \\ \gamma(u) \geq k_1 u & \text{for } u \geq A \\ \psi(w) \geq k_2^{-1} w & \text{for } w > \max(B_1, B_2) \end{cases} \tag{4.25}$$

Using (4.25) one can check that if (4.21) is satisfied and M_* is defined by (4.22), then the inequalities (4.8a) - (4.8c) are satisfied. We know that

$$\gamma(M_* - B_1) \geq k_1(M_* - B_1)$$

so Lemma 4.2 now follows from Lemma 4.1.

We want to use the inner disc lemma and some phase plane ideas from [20], [30] and [44] in order to obtain a priori bounds for slowly oscillating periodic solutions of (4.1). The basic result we shall need is the following generalization of Theorem 4.1 in [30].

LEMMA 4.3. (See Theorem 2.1 of [30]). Let $g: R \to R$ be a continuously differentiable function such that $g(0) = 0$ and g is strictly monotonic increasing and suppose that $x(t)$ is a slowly oscillating periodic solution of

$$\dot{x}(t) = -g(x(t-1))$$

Suppose that $f: R \to R$ is a continuously differentiable function such that $f(0) = 0$ and f is strictly monotonic increasing. Assume that either (A) $|g(x)| < |f(x)|$ for all nonzero x and $g'(g^{-1}(v)) \leq f'(f^{-1}(v))$ for all v in the range of g or (B) $|f(x)| < |g(x)|$ for all nonzero x and $f'(f^{-1}(v)) \leq g'(g^{-1}(v))$ for all v in the range of f. Assume that $y(t)$ is a slowly oscillating periodic solution of

$$\dot{y}(t) = -f(y(t-1))$$

such that $y(t) > 0$ for $0 < t < 2$, $y(t) < 0$ for $2 < t < 4$ and

$y(t+4) = y(t)$ for all t. Write $Y(t) = (y(t), \dot{y}(t))$ and $X(t) = (x(t), \dot{x}(t))$ and let C_Y and C_X denote the simple closed curves in R^2 given by $Y(t)$ and $X(t)$ respectively. Then if assumption A holds and C_X is contained in the closure of the interior of C_Y, C_X and C_Y have no points in common. If assumption B holds and C_Y is contained in the closure of the interior of C_X, C_Y and C_X have no points in common.

REMARK 4.1. In the statement of Theorem 2.1 in [30], the condition that $f(x)$ be strictly monotonic increasing was accidentally omitted; however, this assumption is necessary.

If we assume that $f'(x)$ is nonincreasing on $[0,\infty]$ and nondecreasing on $[-\infty,0]$, which is a hypothesis of Theorem 2.1 in [30], then the assumption that $0 < g(x) < f(x)$ made in [30] implies that

$$g'(g^{-1}(v)) \le f'(f^{-1}(v))$$

for all v in the range of g. A similar conclusion holds if $0 < f'(x) < g'(x)$ for all x. Thus Lemma 4.3 generalizes Theorem 2.1 in [30] and in fact this generalization will be essential for our application.

REMARK 4.2. We do not assume that $f(x)$ is odd, as in [30]. However, this is a dubious generalization, since a periodic solution like $y(t)$ is only known to exist if $f(x)$ is odd and satisfies certain other conditions.

PROOF of LEMMA 4.3. The proof follows the outlines of the proof of Theorem 2.1 in [30]. We shall consider case B, since the proof in case A is analogous. Suppose that z_1 and z_2 are the first and second zeros of $x(t)$. The portions of C_X and C_Y with nonnegative second coordinates are given respectively by $X(t)$ for $z_1 + 1 \le t \le z_2 + 1$ and by $Y(t)$ for $3 \le t \le 5$.

Assume now that C_Y is contained in the closure of the interior of C_X and that C_Y and C_X have a point P in common. We shall obtain a contradiction. Just as in [30], we can assume that $P = y(T_1) = x(t_1)$, where $3 < T_1 \le 5$ and $z_1 + 1 < t_1 \le z_2 + 1$. Define a number τ, with $z_1 + 1 \le \tau < z_2$, by $x(\tau) = y(3)$ and define σ, with $z_2 < \sigma \le z_2 + 1$, by $x(\sigma) = y(5)$. For $\tau \le t \le \sigma$, define $T(t)$ to be the unique number between 3 and 5 such that

$$y(T(t)) = x(t)$$

In other words, $T(t) = u^{-1}(x(t))$, where $u = y|[3,5]$, so that $T(t)$ is continuous on $[\tau,\sigma]$ and differentiable on (τ,σ).

We consider two cases.

Case 1. Assume that $X(t) \notin C_Y$ for $z_1 + 1 < t < z_1 + 2$. Define t_1 to be the first time $t \geq z_1 + 2$ such that $X(t) \in C_Y$, so $z_1 + 2 \leq t_1 \leq z_2 + 1$, and define $T_1 = T(t_1)$. If $\tau < t < t_1$ we have (because $y(T(t)) = x(t)$ for $\tau < t < t_1$)

(4.26) $$y'(T(t))T'(t) = x'(t)$$

Because C_Y is inside C_X, we conclude from (4.26) that

(4.27) $$T'(t) > 1 \quad \text{for} \quad \tau < t < t_1$$

Since $Y(T_1) = X(t_1)$, we have $\dot{y}(T_1) = \dot{x}(t_1)$, so

(4.28) $$f(y(T_1-1)) = g(x(t_1-1)) .$$

Inequality (4.28) implies that

(4.29) $$y(T_1-1) \leq x(t_1-1) \leq 0$$

Because we are assuming that $t_1 - 1 > z_1 + 1$ and because $x(t)$ is increasing for $z_1 + 1 \leq t \leq z_2 + 1$, we conclude from (4.29) that

(4.30) $$\tau \leq t_1 - 1$$

Using (4.30) and (4.27) we obtain

(4.31) $$1 = \int_{t_1-1}^{t_1} dt < \int_{t_1-1}^{t_1} \dot{T}(t)dt = T_1 - T(t_1-1)$$

and (4.31) implies

(4.32) $$y(T(t_1-1)) = x(t_1-1) < y(T_1-1)$$

Inequalities (4.29) and (4.32) are contradictory.

Case 2. Assume that $X(t_1) \in C_Y$ for some t_1 with $z_1 + 1 < t_1 < z_1 + 2$. In this case T is differentiable at t_1 and (2.26) implies that $\dot{T}(t_1) = 1$. If we define $\phi(t)$ by

$$\phi(t) = \dot{y}(T(t)) - \dot{x}(t)$$

$\phi(t)$ is C^1 for t near t_1 (because f and g are C^1) and $\phi(t)$ has a local minimum at t_1 (because C_Y is inside C_X), so $\dot{\phi}(t_1) = 0$. Computing $\dot{\phi}(t_1)$ gives

(4.33) $$-f'(y(T_1-1))\dot{y}(T_1-1) + g'(x(t_1-1))\dot{x}(t_1-1) = 0$$

We know that $\dot{x}(t_1-1) < 0$ because $z_1 < t_1-1 < z_1 + 1$. Therefore we must have $y'(T_1-1) < 0$ (otherwise (4.33) gives an immediate contradiction), and we conclude that $2 < T_1-1 < 3$.

Just as in case 1, equations (4.28) and (4.29) are valid, and because $y(2) = 0$, the intermediate value theorem implies that there exists a number τ_* with

(4.34) $$2 < \tau_* < T_1-1 < 3$$

such that

(4.35) $$y(\tau_*) = x(t_1-1)$$

Equality (4.28) and the fact that C_Y is inside C_X imply that

(4.36) $$\dot{x}(t_1-1) \le \dot{y}(\tau_*) < 0$$

On the other hand, by using the defining equation for $y(t)$ and the fact that y is decreasing on $[1,3]$, one finds that

(4.37) $$\dot{y}(\tau_*) < \dot{y}(T_1-1) < 0$$

Inequalities (4.36) and (4.37) imply that

(4.38) $$|\dot{y}(T_1-1)| < |\dot{x}(t_1-1)|$$

Furthermore, because $f(y(T_1-1)) = g(x(t_1-1))$, the assumptions of our lemma imply that

(4.39) $$0 < f'(y(T_1-1)) \le g'(x(t_1-1))$$

Inequalities (4.38) and (4.39) contradict (4.33). Since we have obtained a contradiction in all cases, the lemma is proved.

<u>REMARK 4.3</u>. Suppose that f and g are C^1 functions such that $f'(x) > 0$ for all nonzero x, $g'(x) > 0$ for all nonzero x and $f(0) = g(0) = 0$. In addition suppose that

(4.40) $$f'(f^{-1}(x)) \le g'(g^{-1}(x))$$

for all x in the range of f and that the inequality in (4.40) is strict for $0 < |x| \le \delta$, δ some positive number. Inequality (4.40) implies that

(4.41) $$(f^{-1})'(x) \geq (g^{-1})'(x)$$

for all x in range f, with strict inequality for $0 < |x| \leq \delta$. We obtain from (4.41) that

$$|f^{-1}(x)| > |g^{-1}(x)|$$

for all nonzero x in the range of f and consequently that

(4.42) $$|g(x)| > |f(x)|$$

for all nonzero x. It follows that one of the assumptions in case B of Lemma 4.2 is almost redundant. A similar observation holds for case A.

REMARK 4.4. Before proving our first theorem we also need to recall some results from [19], [29] and [30] about special periodic solutions of (4.1). Let $f(x)$ be a continuous, odd function such that $xf(x) > 0$ for nonzero x. Suppose that either (1) $f'(0) < (\frac{\pi}{2})$ and $\lim_{x\to\infty} x^{-1}f(x) > (\frac{\pi}{2})$ or (2) $f'(0) > (\frac{\pi}{2})$ and $\lim_{x\to\infty} x^{-1}f(x) < (\frac{\pi}{2})$ (it is understood that all limits exist, although $+\infty$ is allowed as a value in case (1)). Then (see [19] the equation

(4.43) $$\dot{x}(t) = -f(x(t-1))$$

has a periodic solution such that

(a) $x(t) > 0$ for $0 < t < 2$, (b) $x(t+2) = -x(t)$ for all t and (c) $x(-t) = -x(t)$ for all t. In the terminology of [29], if Z denotes the Banach space of continuous functions $x(t)$ which satisfy (b) and (c) (in the supnorm) and $K = \{x\in Z : x(t) \geq 0 \text{ for } 0 \leq t \leq 2\}$, equation (4.43) has a solution in $K-\{0\}$. If either $x^{-1}f(x) < (\frac{\pi}{2})$ for $0 < x \leq \beta$ or $x^{-1}f(x) > (\frac{\pi}{2})$ for $0 < x \leq \beta$, the arguments in [29] show that any solution $x \in K - \{0\}$ of (4.1) must satisfy $\|x\| = \sup_t x(t) > \beta$. If f is odd and continuous with $f(x)>0$ for $x>0$ and either $x^{-1}f(x)$ is strictly increasing or strictly decreasing for $x>0$, Theorem 1.3 of [30] implies that equation (4.1) has a most one solution $x \in K - \{0\}$.

We shall also need information about the case when f depends on a parameter α. Assume that, for $\alpha_1 \leq \alpha \leq \alpha_2$, $f_\alpha(x) \equiv f(x,\alpha)$ is an odd function which is positive for x positive and that $f(x,\alpha)$ is continuous. It is known (see [19] that solutions $x \in K - \{0\}$ of

(4.44) $$\dot{x}(t) = -f_\alpha(x(t-1))$$

are in one-one correspondence with solutions of minimal period 4 of the initial value problem

$$(4.45)\qquad \begin{cases} \dot{x}(t) = -f_\alpha(y(t)) \\ \dot{y}(t) = f_\alpha(x(t)) \\ x(1) = c > 0\ ,\ y(1) = 0 \end{cases}$$

If $T(c,\alpha)$ denotes the minimal time $t > 0$ such that $x(1+t) = 0$ (where $(x(t),y(t))$ solves (4.45), then $(x(t),y(t))$ is periodic of period $4T(c,\alpha)$. Thus the set of pairs $(x,\alpha) \in (K-\{0\} \times [\alpha_1,\alpha_2]$ such that $x(t)$ satisfies (4.45) is homeomorphic to $\{(c,\alpha)\,|\,c>0\ ,\ \alpha \in [\alpha_1,\alpha_2]$ and $T(c,\alpha) = 1\}$.

Now assume that there are nonnegative numbers $b_1 < \frac{\pi}{2}$ and $b_2 > \frac{\pi}{2}$ and positive numbers r_1 and r_2 (with $r_2 > r_1$) such that either (1) $x^{-1}f_\alpha(x) \le b_1$ for $|x| \le r_1$ and $x^{-1}f_\alpha(x) \ge b_2$ for $|x| \ge r_2$ (for all α) or (2) $x^{-1}f_\alpha(x) \ge b_2$ for $|x| \le r$, and $x^{-1}f_\alpha(x) \le b_1$ for $|x| \ge r_2$. It is easy to see (using arguments as in [19]) that there exist positive numbers r and R , $r < R$, such that, in case 1 above, $T(R,\alpha) < 1$ and $T(r,\alpha) > 1$ for $\alpha_1 \le \alpha \le \alpha_2$ while in case 2, $T(R,\alpha)<1$ and $T(r,\alpha) < 1$ for $\alpha_1 \le \alpha \le \alpha_2$. It now follows that there exists a connected set $S \subset (r,R) \times [\alpha_1,\alpha_2]$ such that $T(c,\alpha) = 1$ for all (c,α) in S and S has nonempty intersection with $(r,R) \times \{\alpha\}$ for all α in $[\alpha_1,\alpha_2]$. This fact is actually a special case of a much more general result which can be proved with degree theory and which goes back to work of Leray and Schauder. A general version is proved in Lemma 3.4 of [30]. The existence of S of course implies the existence of a corresponding continuum of solutions of (4.44) in $(K-\{0\}) \times [\alpha_1,\alpha_2]$.

We are now in a position to prove our first theorem.

THEOREM 4.1. Let Λ be a compact metric space and $g\colon R \times \Lambda \to R$ be a continuous map such that $g(0,\lambda) = 0$ for all $\lambda \in \Lambda$. Write $g_\lambda(x)$ for $g(x,\lambda)$ and assumme that $\frac{d}{dx}g_\lambda(x) = g_\lambda'(x)$ exists for all $(x,\lambda) \in R \times \Lambda$, $g_\lambda'(x) > 0$ for all nonzeros x and $(\lambda,x) \to g_\lambda'(x)$ is continuous. Assume either (1) there exist constants $k_1 < (\frac{\pi}{2})$ and $A > 0$ such that $g_\lambda'(x) \le k_1$ for all $\lambda \in \Lambda$ and all x such that $|x| \ge A$ or (2) there exist constants $k_1 > (\frac{\pi}{2})$ and $A > 0$ such that $g_\lambda'(x) \ge k_1$ for all $\lambda \in \Lambda$ and all x with $|x| \ge A$. In case (2) let $g_\lambda^{-1}(y)$ denote the inverse function of the map $x \to g_\lambda(x)$ and assume that there exists a C^1 function $g(x)$ with $g(0) = 0$ such that $g'(x) > 0$ for all nonzero x , the range of g is all of R , and

$$(4.46)\qquad (g^{-1})'(x) \ge (g_\lambda^{-1})'(x)$$

for all nonzero x and all $\lambda \in \Lambda$. Then there exists a constant M such that if $x(t)$ is a slowly oscillating periodic solution of

$$\dot{x}(t) = -g_\lambda(x(t-1))$$

for some $\lambda \in \Lambda$, one has

$$\sup_t |x(t)| \le M$$

PROOF. We consider case (2) first. Our first objective is to construct a parametrized family of C^1 odd functions f_α for $\alpha \ge 0$ such that for all $\lambda \in \Lambda$ and nonzero x

$$f'_\alpha(f_\alpha^{-1}(x)) < g'_\lambda(g_\lambda^{-1}(x)) \tag{4.47}$$

and such that certain further conditions are satisfied. Let c be a positive constant with $c < \frac{\pi}{2}$ and for each $\alpha \ge 0$ define a C^1 odd function $h_\alpha(x)$ by $h_\alpha(0) = 0$ and

$$h'_\alpha(x) = \begin{cases} c & \text{for } |x| \le \alpha+A \\ (\alpha+A+1-|x|)c + (|x|-\alpha-A)k_1 & \text{for } \alpha+A \le |x| \le \alpha+A+1 \\ k_1 & \text{for } |x| \ge \alpha+A+1 \end{cases}$$

Next we need to modify the function g whose existence has been assumed. First define a constant B by

$$B = \sup \{|g_\lambda(x)| : |x| \le A , \lambda \in \Lambda\}$$

It is not hard to see that

$$(g_\lambda^{-1})'(y) \le k_1^{-1} \quad \text{for } |y| \ge B \tag{4.48}$$

By replacing g^{-1} by cg^{-1} , where $c \ge 1$, we can preserve (4.46) and also insure that

$$(g^{-1})'(x) \ge k_1^{-1} , \quad x = \pm B$$

Now define the inverse j^{-1} of a C^1 function j by $j^{-1}(y) = g^{-1}(y)$ for $|y| \le B$ and

$$(j^{-1})'(y) = \begin{cases} k_1^{-1} \quad \text{for } |y| \ge B+1 \\ (B+1-|y|)(g^{-1})'((\operatorname{sgn} y)B)+(|y|-B)k_1 , \\ \qquad\qquad \text{for } B \le |y| \le B+1 \end{cases} \tag{4.49}$$

One can check that j satisfies the same conditions as g and that, in addition,

$$(j^{-1})'(y) = k_1^{-1} \quad \text{for } |y| \ge B+1 \tag{4.50}$$

We want to modify j so that it will be odd and still satisfy (4.46),

(4.48), etc. For this purpose define $\beta(t)$ by

$$\beta(t) \equiv \max\,((j^{-1})'(t)\;,\;(j^{-1})'(-t))$$

Notice that $\beta(t)$ is continuous (except possibly at 0), and that it has an improper Riemann integral on any interval $[0,x]$ because

$$(4.51)\qquad \beta(t) \le (j^{-1})'(t) + (j^{-1})'(-t)$$

and because the right hand side of (4.51) has an improper Riemann integral on $[0,x]$. Therefore we can define the inverse k^{-1} of a C^1 function k by

$$k^{-1}(x) \equiv \int_0^x \beta(t)\,dt$$

One easily checks that k is odd, $k'(x) > 0$ for $x \neq 0$, and the range of k is all of R. Furthermore k satisfies (4.46) (substituting k for g) and (4.50) (substituting k for j).

We can now define f_α by setting $f_\alpha(0) = 0$ and defining the inverse function f_α^{-1} by

$$(f_\alpha^{-1})'(y) = \delta + \max\,((j^{-1})'(y),\;(h_\alpha^{-1})'(y))$$

where δ is a positive constant such that $\delta + k_1^{-1} < (\frac{2}{\pi})$. It is not hard to see that f_α is a C^1 odd function which satisfies (4.47), that $f_\alpha'(x) \le c$ for $|x| \le \alpha + A$ and that $f_\alpha'(x) \ge (\delta + k_1^{-1})^{-1}$ for $|x|$ sufficiently large.

Remark 4.4 now shows that there exists a constant R_0 such that any special periodic solution $y(t)$, $y \in K-\{0\}$, of

$$\dot{y}(t) = -\,f_0(y(t-1))$$

satisfies

$$(y(t))^2 + (\dot{y}(t))^2 < R_0^2$$

We next claim that the inner disc lemma implies the existence of a constant M such that if $x(t)$ is a slowly oscillating periodic solution of

$$(4.52)\qquad x'(t) = -\,g_\lambda(x(t-1))$$

for some $\lambda \in \Lambda$ and $\|x\| \ge M$, then

$$(4.53)\qquad (x(t))^2 + (\dot{x}(t))^2 \ge R_0^2$$

for all t . To see this define for each $\lambda \in \Lambda$ functions h_λ, γ_λ and ψ_λ as given by equation (4.7) with g_λ substituted for g . Thus one has for example

$$h_\lambda(u) \equiv \max \{|g_\lambda(x)| : |x| \le u\}$$

Next define functions h, γ and g by

$$(4.54) \qquad \begin{cases} h(u) = \max \{|g(x,\lambda)| : |x| \le u, \lambda \in \Lambda\} \\ \gamma(u) = \inf \{|g(x,\lambda)| : |x| \ge u, \lambda \in \Lambda\} \\ \psi(w) = \inf \{ u \ge 0 : h(u) = w \} \end{cases}$$

It is trivial to check that $h(u) \ge h_\lambda(u)$, $\gamma(u) \le \gamma_\lambda(u)$, $\psi(w) \le \psi_\lambda(w)$, $\lim_{w\to+\infty} \psi(w) = +\infty$, and $\gamma(u) \ge k_1 u$ for $u \ge A$. Using this information and Lemma 4.1 one can see that if $M' \equiv \psi(M)$ ($B_1 = 0$ in the notation of Lemma 4.1) and if M_* satisfies (4.8a) - (4.8c) then any solution $x(t)$ of (4.52) such that $\|x\| \ge M$ satisfies

$$(4.55) \qquad (x(t))^2 + (\dot{x}(t))^2 \ge \min ((\gamma(M_*))^2, M_*^2)$$

For M large enough one obtains (4.53). Our only difficulty, of course, was to obtain estimates independent of λ .

With this choice of M , we claim that any slowly oscillating periodic solution of (4.52) for some λ satisfies $\|x\| \le M$. For suppose not and let $x(t)$ be such a solution with $\|x\| > M$. Define $X(t) = (x(t), x'(t)) \in R^2$ and let C_X denote the simple closed curve given by $X(t)$. Select $\alpha_1 > \|x\|$ and by Remark 4.4 let S denote a bounded connected set in $(K-\{0\}) \times [0,\alpha_1]$ of solutions of

$$(4.56) \qquad \dot{y}(t) = - f_\alpha(y(t))$$

That is, $(y,\alpha) \in S$ implies y satisfies (4.56). Also, $S \cap ((K-\{0\}) \times \{\alpha\})$ is nonempty for $0 \le \alpha \le \alpha_1$. For $(y,\alpha) \in S$, let $Y(t) = (y(t), \dot{y}(t))$ and let C_Y denote the corresponding curve in R^2 . Let V_1 denote $\{(y,\alpha) \in S \mid C_Y$ is contained in the closure of the interior of $C_X\}$ and let V_2 denote the complement of V_1 in S , so V_2 is open. Any point $(y,\alpha_1) \in S$ satisfies $\|y\| > M$, so V_2 is nonempty, and by construction all points $(y,0) \in S$ lie in V_1 , so V_1 is nonempty. Inequality (4.47) and Remark 4.3 imply that the hypotheses of Lemma 4.3 are satisfied, so if $(y,\alpha) \in V_1$, the curve C_Y is contained in the interior of C_X . Using this fact one can see that V_1 is open in the relative topology on S . We now have a contradiction: the connected topological space S is the disjoint union of two nonempty open sets.

It remains to consider case 1 of the theorem. For technical reasons the proof in this case is analogous to the proof for case 2 but much simpler. Define a constant B by

$$B = \sup \{g'_\lambda(x) \mid \lambda \in \Lambda\ ,\ |x| \le A\} + (\tfrac{\pi}{2})$$

and define a parametrized family of odd functions $f_\alpha(x)$ (unrelated to the functions in case 2) by $f_\alpha(0) = 0$ and

$$f'_\alpha(x) = \begin{cases} B & \text{for } |x| \le A+\alpha \\ B(A+\alpha+1-|x|) + (|x|-\alpha-A)k'_1 & \text{for } A+\alpha \le |x| \le A+\alpha+1 \\ k'_1 & \text{for } |x| \ge A+\alpha+1 \end{cases}$$

where $k_1 < k'_1 < (\frac{\pi}{2})$. Theorem 1.3 of [30] implies that the equation

$$\dot{y}(t) = -\ f_\alpha(y(t-1))$$

has a unique solution $y(t) = y_\alpha(t)$ such that $y(0) = 0$, $y(t) > 0$ for $0 < t < 2$, $y(t+2) = -\ y(t)$ and $y(-t) = -\ y(t)$ for all t . Furthermore, Theorem 1.3 in [30] implies that $\alpha \to y_\alpha \in C[0,4]$ is continuous and that $\sup_t |y_\alpha(t)| > A + \alpha$. As usual, define

$Y_\alpha(t) = (y_\alpha(t), \dot{y}_\alpha(t))$ and let C_{Y_α} be the curve given by $Y_\alpha(t)$.

We claim that if $x(t)$ is any slowly oscillating periodic solution of (4.52), then C_X is contained in the interior of C_{Y_0} . To see this notice that because $f'_\alpha(x) > g'_\lambda(x)$ for all x and $f'_\alpha(x)$ is a nonincreasing function of x , it is easy to check that the hypotheses of case A of Lemma 4.3 are satisfied for $f = f_\alpha$ and $g = g_\lambda$. It follows that if C_X is contained in the closure of the interior of C_{Y_α} for some α then C_X and C_{Y_α} are disjoint. Suppose by way of contradiction that C_X is not contained in the closure of the interior of C_{Y_0} . Lemma 4.2 and the fact that $\sup_t |y_\alpha(t)| > A+\alpha$ imply that C_X is contained in the interior of C_{Y_γ} for some $\gamma > 0$. Define β by

$$\beta = \inf \{\alpha \ge 0 \mid C_X \text{ is contained in the interior of } C_{Y_\alpha}\}$$

Our assumption insures that $\beta > 0$, that C_X is contained in the closure of the interior of C_{Y_β} and that $C_X \cap C_{Y_\beta}$ is nonempty, a contradiction.

We conclude that C_X is contained in the closure of the interior of C_{Y_0} , which implies that

$$\sup_t |x(t)| \le M \quad \sup_t |y_o(t)|$$

The latter estimate completes the proof.

REMARK 4.5. Under the assumptions of the first two sentences of Theorem 4.1 one can prove that $(x,\lambda) \to g_\lambda^{-1}(x)$ is continuous for $(x,\lambda) \in R \times \Lambda$ and that $(x,\lambda) \to (g_\lambda^{-1})'(x)$ is continuous for $x \neq 0$. If one defines a continuous function $\gamma(x)$ (for $x \neq 0$) by

$$\gamma(x) = \max_{\lambda \in \Lambda} (g_\lambda^{-1})'(x)$$

and if one assumes that $\gamma(x)$ has an improper Riemann integral on $[0,a]$ and $[-a,0]$ for some positive a, then one can define g^{-1}, the inverse function of a C^1map g, by

$$g^{-1}(x) = \int_0^x \gamma(t)dt$$

and $g(x)$ satisfies all the conditions of Theorem 4.4. Conversely, if one assumes the existence of g as in the statement of Theorem 4.1, then $\gamma(x)$ has an inproper Riemann integral.

We wish to show now how Theorem 4.1 can be used to obtain new existence theorems about periodic solutions of (4.1). First we need to recall some standard results. Let K denote the cone of nondecreasing, continuous functions ϕ in $C[0,1]$ such that $\phi(0) = 0$. Let $g: R \to R$ denote a given continuous function such that $xg(x) > 0$ for $x \neq 0$. For $\varphi \in K$ consider the equation

$$\text{(4.57)} \qquad \begin{cases} \dot{x}(t) = -g(x(t-1)), & t \ge 1 \\ x|[0,1] = \varphi \end{cases}$$

and let $x(t;\varphi) = x(t)$ denote the unique solution of (2.57). If φ is not identically zero, let $z_1 = z_1(\varphi)$ denote the first time $t > 1$ such that $x(t) = 0$ and $z_2 = z_2(\varphi)$ denote the first time $s > z_1(\varphi)$ such that $x(s) = 0$. If z_1 or z_2 is not defined, write $z_2(\varphi) = \infty$. Define a map $F: K \to K$ by $(F\varphi)(t) = x(z_2+t)$ if $z_2 = z_2(\varphi) < \infty$ and $F(\varphi) = 0$ if $z_2 = \infty$ or $\varphi = 0$. It is known [26,27] that F is a compact continuous map of K into K whose nonzero fixed points give slowly oscillating periodic solutions. Furthermore, suppose $g'(0)$ is defined and unequal to $\frac{\pi}{2}$. Then there is an $\varepsilon > 0$ such that $F(\varphi) \neq \varphi$ for $\varphi \in K$ and $\|\varphi\| = \varepsilon$; and if $V = \{\varphi \in K : \|\varphi\| < \varepsilon\}$ and $i_K(F,V)$ denotes the fixed point index of $F: V \to K$, then $i_K(F,V) = 1$ if $0 \le g'(0) < (\frac{\pi}{2})$ and $i_K(F,V) = 0$ if $g'(0) > (\frac{\pi}{2})$.

THEOREM 4.2. Let $g: R \to R$ be a continuously differentiable function such that $g'(x) > 0$ for all nonzero x and $g(0) = 0$. Assume either (1) $g'(0) > (\frac{\pi}{2})$ and $\limsup_{|x| \to \infty} g'(x) < (\frac{\pi}{2})$ or (2) $g'(0) < (\frac{\pi}{2})$ and $\liminf_{|x| \to \infty} g'(x) > (\frac{\pi}{2})$. Then the equation

$$\dot{x}(t) = -g(x(t-1)) \tag{4.58}$$

has a slowly oscillating periodic solution, and there exists a constant M such that $\sup_t |x(t)| \leq M$ for any slowly oscillating periodic solution of (4.58).

PROOF. Define $g_1(x) = kx$, where $0 < k < (\frac{\pi}{2})$ if g satisfies case 1 and $k > (\frac{\pi}{2})$ if g satisfies case 2. Define $g_\lambda(x)$ for $0 \leq \lambda \leq 1$ by

$$g_\lambda(x) = (1-\lambda)g(x) + \lambda g_1(x)$$

Using Remark 4.5 one can check that the conditions of Theorem 4.1 are satisfied, so there exists a constant M such that if $x(t)$ is a slowly oscillating periodic solution of

$$\dot{x}(t) = -g_\lambda(x(t-1)) \tag{4.59}$$

for some λ with $0 \leq \lambda \leq 1$, then

$$\sup_t |x(t)| < M \tag{4.60}$$

Let $F_\lambda : K \to K$ be the map (corresponding to equation (4.59)) defined as in the paragraph before Theorem 4.2. Inequality (4.60) shows that if $W = \{\varphi \in K \mid \|\varphi\| < M\}$, then $i_K(F_\lambda, W)$ is defined for $0 \leq \lambda \leq 1$; and the homotopy property for the fixed point index implies that

$$i_K(F_1, W) = i_K(F_0, W) \tag{4.61}$$

By the remarks immediately preceding Theorem 4.2, we know that $i_K(F_1, W) = 1$ if g satisfies case 1 and $i_K(F_1, W) = 0$ in case 2. This gives

$$\begin{cases} i_K(F_0, W) = 1 & \text{case 1} \\ i_K(F_0, W) = 0 & \text{in case 2} \end{cases} \tag{4.62}$$

On the other hand, we know there exists $\varepsilon > 0$ such that $F_0(\varphi) \neq \varphi$ for $\|\varphi\| = \varepsilon$ and if $U = \{\varphi \in K : \|\varphi\| < \varepsilon\}$,

$$\begin{cases} i_K(F_0, U) = 0 & \text{in case 1} \\ i_K(F_0, U) = 1 & \text{in case 2} \end{cases} \tag{4.63}$$

f we define $V = \{\varphi \in K : \varepsilon < \|\varphi\| < M\}$, the additivity property implies

$$(4.64)\qquad i_K(F_o,V) = i_K(F_o,W) - i_K(F_o,U)$$

so we obtain

$$(4.65)\qquad \begin{cases} i_K(F_o,V) = 1 \\ i_K(F_o,V) = -1 \end{cases}$$

In either case, the fixed point index of F_o on V is nonzero, so F_o has a fixed point in V.

REMARK 4.6. A version of Theorem 4.2 can be obtain for functions g which are piecewise C^1. It also seems likely that the ideas of this section can be generalized to equations

$$\dot{x}(t) = -g(x(t),x(t-1))$$

For reasons of length we do not consider these questions.

REMARK 4.7. Theorem 4.2 should be compared with Theorem 2.1 of [26]. If one restricts attention to C^1 functions g such that $g'(x) > 0$ for $x \neq 0$ and $\lim_{x\to+\infty} g'(x) = c_+$ and $\lim_{x\to-\infty} g'(x) = c_-$ (where we allow $c_+ = +\infty$ or $c_- = +\infty$), then Theorem 2.1 of [26] only applies if $c_+ = c_- > (\frac{\pi}{2})$, while our Theorem 4.2 only requires $c_+ > (\frac{\pi}{2})$ and $c_- > (\frac{\pi}{2})$. Thus our Theorem 4.2 provides a partial answer to a conjecture raised in Remark 2.2 of [26] by removing a condition of "asymptotic oddness at ∞".

REMARK 4.8. One can also consider a parametrized family of equations

$$\dot{x}(t) = -g_\alpha(x(t-1)),\quad \alpha > 0$$

where the g_α satisfy conditions as in Theorem 4.2. One can obtain global continua of slowly oscillating periodic solutions as in [28] by using Theorem 1.1 of [28]. Again, for reasons of length we give no details.

The a priori bounds of Theorem 4.1 require that the nonlinearity $g(x)$ satisfy $xg(x) > 0$ for $x \neq 0$. As our final result we wish to show that this condition is unnecessary if $\lim_{|x|\to\infty} (\frac{g(x)}{x}) = \beta > (\frac{\pi}{2})$. The proof consists of a refinement of the argument in Lemma 2.2 of [26]

THEOREM 4.3. (Compare Lemma 2.2 in [26]). Assume that $g: R \to R$ is a continuous map and that

$$(4.66)\qquad \lim_{|x|\to\infty} (\frac{g(x)}{x}) = \beta$$

where $(\frac{\pi}{2}) < \beta < \infty$. Then there exists a number M_o such that any slowly oscillating periodic solution of

$$\dot{x}(t) = -g(x(t-1)) \tag{4.67}$$

satisfies $\sup_t |x(t)| \le M_o$.

REMARK 4.9. Arguments like those in [26] also prove Theorem 4.3 when $\beta = +\infty$ (assuming g is nondecreasing on intervals $(-\infty,-A]$ and $[A,\infty)$, for some A). We omit the proof.

PROOF of Theorem 4.3. Suppose that $x(t)$ is a slowly oscillating periodic solution of (4.67) and that $\sup_t |x(t)| = M$. If z_1 and z_2 are the first and second zeros of $x(t)$, we know that $|x(t)|$ assumes the value of M on either $(0,z_1)$ or (z_1,z_2) . We claim we can assume $x(t_1)=M$ for some $t_1 \in (0,z_1)$. If not, define $x_1(t) = -x(z_1+t)$ and $g_1(x) = -g(-x)$. Then $x_1(t)$ is a slowly oscillating periodic solution of $\dot{x}_1(t) = -g_1(x(t-1))$, $x_1(t) = M$ for some $t \in (0,z_2-z_1)$, and we could apply the following argument to $x_1(t)$.

Therefore we assume that $x(t) = M$ for some $t \in (0,z_1)$. Define constants B_1 and A such that

$$\begin{cases} g(x) > -B_1 & \text{for } x \ge 0 \\ g(x) < B_1 & \text{for } x \le 0 \\ xg(x) > 0 & \text{for } |x| \ge A \end{cases}$$

Then the same argument used in the inner disc lemma shows that if $M > B_1 + A$

$$x(1) \ge M - B_1$$

If we define $y(t) = x(t+1)$, then $y(-1) = 0$ and $y(t)$ satisfies (4.67). Recall (see [27] that the characteristic equation

$$\lambda + \beta e^{-\lambda} = 0$$

has a unique solution $\lambda = \mu + i\nu$ such that $\mu > 0$ and $(\frac{\pi}{2}) < \nu < \pi$. Equation (2.4) in [26] gives

$$\begin{cases} -\lambda \int_{-1}^{0} y(t)e^{-\lambda t}dt - y(0) = N \\ N = \int_{0}^{\infty} [-f(y(t-1)) + \beta y(t-1)]e^{-\lambda t}dt \end{cases} \tag{4.68}$$

If we use the fact that $-\lambda = \beta e^{-\lambda}$ in (4.68) and take real and imaginary parts of (4.68) we obtain

(4.69) $$y(0)-\beta \int_{-1}^{0} y(t)e^{-\mu(t+1)} \cos \nu(t+1)dt = - \mathrm{Re}(N)$$

(4.70) $$\beta \int_{-1}^{0} y(t)e^{-\mu(t+1)} \sin \nu(t+1)dt = \mathrm{Im}(N)$$

For reasons that will become apparent, define a positive constant k by

$$k = e^{-\mu} \min(\sin(\tfrac{1}{2}\nu),\sin(\nu))$$

and define $c = (2+k)^{-1}$, so $(1-2c) = ck$. We consider two cases: $\beta \int_{-\frac{1}{2}}^{0} y(t)dt \le cM$ (case 1) or not (case 2).

In case 1 we obtain

(4.71) $$\begin{cases} -\beta \int_{-(\frac{1}{2})}^{0} y(t)e^{-\mu(t+1)} \cos \nu(t+1)dt \ge -\beta\int_{-(\frac{1}{2})}^{0} y(t)dt \ge - cM \\ \text{and } -\beta \int_{-1}^{-(\frac{1}{2})} y(t)e^{-\mu(t+1)} \cos \nu(t+1)dt \ge -\beta\int_{-1}^{-(\frac{1}{2})} y(t)dt \end{cases}$$

Since $\dot{y}(t) \ge - B_1$ for $-1 \le t \le 0$, we have

$$y(t) \le y(t+\tfrac{1}{2}) + (\tfrac{1}{2})B_1 , \quad -1 \le t \le -\tfrac{1}{2}$$

and

(4.72) $$\begin{cases} -\beta \int_{-1}^{-(\frac{1}{2})} y(t)dt \ge -\beta\int_{-1}^{-\frac{1}{2}} [y(t+\tfrac{1}{2}) + (\tfrac{1}{2})B_1]dt \\ \qquad \ge -(\tfrac{\beta}{4})B_1 - cM \end{cases}$$

Equations (4.71) and (4.72) together imply

(4.73) $$-\beta \int_{-1}^{0} y(t)e^{-\mu(t+1)} \cos \nu(t+1)dt \ge -2cM - (\tfrac{\beta}{4})B$$

By substituting (4.73) in (4.69) and recalling that $y(0) \ge M-B_1$ we find that

(4.74) $$(1-2c)M \le |N| + B_1 + (\tfrac{\beta}{4})B_1$$

Next suppose that case 2 holds. Then because $y(t)$ and $\sin \nu(t+1)$ are nonnegative for $-1 \le t \le 0$ we obtain

(4.75) $$\begin{cases} \beta \int_{-1}^{0} y(t)e^{-\mu(t+1)} \sin \nu(t+1)dt \le \beta k \int_{-\frac{1}{2}}^{0} y(t)dt \\ \qquad \le ckM \end{cases}$$

Using (4.75) in (4.70) gives

(4.76) $ckM \le |N|$

It follows from (4.74) and (4.76) that in any case we have

(4.77) $ckM \le |N| + B_1 + (\frac{\beta}{4})B_1$

Select a positive number ε such that

$$\varepsilon\mu^{-1} = (\tfrac{1}{2})ck$$

and, using the hypothesis on f , select M_1 such that

$$|f(x) - \beta x| < \varepsilon\, |x| \quad \text{for} \quad |x| \ge M_1$$

If C is a constant such that

$$|f(x) - \beta x| \le C \qquad \text{for} \quad |x| \le M_1$$

then obvious estimates show

$$|N| \le \int_0^\infty (\varepsilon M + C)e^{-\mu t}dt = \varepsilon\mu^{-1}M + C\mu^{-1}.$$

Using (4.78) in (4.77) gives

(4.79) $(\frac{kc}{2})M \le C\mu^{-1} + B_1 + (\frac{\beta}{4})B_1$

and (4.79) provides an upper bound for M .

REFERENCES

[1] ALLGOWER, E. L., GLASHOFF, K. and PEITGEN, H. O. (eds.): Numerical Solution of Nonlinear Equations, Springer Lecture Notes in Mathematics 878, (1981).

[2] AMANN, H.: Fixed point equations and nonlinear eigenvalue problems in ordered Banach spaces, SIAM Review 18,(1976), 620-709.

[3] ARNOLD, V. I.: Ordinary Differential Equations, MIT Press Cambridge, Massachusetts, (1973).

[4] BIRKHOFF, G. D.: On the periodic motions of dynamical systems, Acta Math. 50, 359-379, (1927).

[5] BIRKHOFF, G.D.: A new criterion of stability, Atti Congr. Intern. d. Matem. Bologna 5, 5-13,(1928).

[6] BOHL, E.: On the bifurcation diagram of discrete analogues for ordinary bifurcation problems, Math. Meth. Appl. Sci. 1,(1979), 566-571.

[7] COLLATZ, L.: The numerical treatment of differential equations, Springer Verlag, Berlin,(1960).

[8] COLLET, P., ECKMANN, J.P. and LANFORD, O. E.III : Universal properties of maps of an interval, Comm. Math. Phys. 76, 211-254,(1980).

[9] COLLET, P. and ECKMANN, J. P.: Iterated Maps on the Interval as Dynamical Systems, Progress in Physics, Vol. 1, Birkhäuser Verlag, Basel and Boston,(1980).

[10] CRANDALL, M. G. and RABINOWITZ, P. H.: Bifurcation from simple eigenvalues, J. Functional Analysis 8,(1971), 321-340.

[11] FEIGENBAUM, M. J.: The universal metric properties of nonlinear transformations, J. Stat. Phys. 21, 669-706,(1979).

[12] GEAR, C. W.: Numerical Initial Value Problems in Ordinary Differential Equations, Prentice-Hall, New Jersey, (1971).

[13] GREENE, J. M.: A method for determining a stochastic transition, J. Math. Phys. 20, 1183-1201,(1979).

[14] GREENE, J. M., MAC KAY, R. S., VIVALDI, F. and FEIGENBAUM, M. J.: Universal behaviour in families of area-preserving maps, Physica 3D, 468-486,(1981).

[15] GROSSMANN, S. and THOMAE, S.: Invariant distributions and stationary correlation functions, Z. Naturforsch. 32, 1353-1363,(1977).

[16] GUMOVSKI, I. and MIRA, C.: Recurrences and Discrete Dynamic Systems, Springer Lecture Notes in Math., Springer Verlag, Berlin, Heidelberg, New York, 809, (1980).

[17] HENRICI, P.: Discrete Variable Methods in Ordinary Differential Equations, John Wiley, New York, (1962).

[18] ISAACSON, E. and KELLER, H.: Analysis of Numerical Methods, John Wiley, New York, (1966).

[19] KAPLAN, J. and YORKE, J.: Ordinary differential equations which yield periodic solutions of differential delay equations, J. Math. Anal. Appl. 48, (1974), 317-324.

[20] KAPLAN, J. and YORKE, J.: On the nonlinear differential delay equation $x'(t) = - f(x(t), x(t-1))$, J. Differential Equations 23, (1977), 293-314.

[21] KRASNOSEL'SKII, M. A.: Positive solutions of Operator Equations, P. Noordhoff, Groningen, (1964).

[22] LASLETT, L. J., MCMILLAN, E. M. and MOSER, J.: Long-term stability for particle orbits, AEC Research and Development Report, Courant Institute of Mathematical Sciences, New York University, (1968).

[23] MEYER, K. R.: Generic bifurcation of periodic points, Trans. Amer. Math. Soc. 149, 95-107, (1970).

[24] MOSER, J.: Stable and Random Motions in Dynamical Systems, Ann. of Math. Studies 77, Princeton Univ. Press, (1973).

[25] NEWHOUSE, S. E.: Lectures on dynamical systems, Dynamical Systems, C.I.M.E. Lectures, Bressanone, Italy, June 1978, Progress in Mathematics, vol. 8, Birkhäuser Verlag, Basel and Boston, pp. 1-114, (1980).

[26] NUSSBAUM, R. D.: Periodic solutions of some nonlinear autonomous F.D.E's. II., J. Differential Equations 14, (1973), 360-394.

[27] NUSSBAUM, R. D.: Periodic solutions of some nonlinear autonomous functional differential equations, Ann. Mat. Pura Appl., 101, (1974), 263-306.

[28] NUSSBAUM, R. D.: A global bifurcation theorem with applications to functional differential equations, J. Functional Analysis 19, (1975), 319-339.

[29] NUSSBAUM, R. D.: Periodic solutions of special differential-delay equations: an example in non-linear functional analysis, Proc. Royal Soc. Edinburgh, 81A, (1978), 131-151.

[30] NUSSBAUM, R. D.: Uniqueness and nonuniqueness for periodic solutions of $x'(t) = -g(x(t-1))$, J. Differential Equations 34, (1979), 25-54.

[31] PEITGEN, H. O.: Phase transitions in the homoclinic regime of area preserving diffeomorphisms, Proc. International Symp. on Synergetics, H. Haken, ed., Springer Series in Synergetics vol.17,(1982), 197-214.

[32] PEITGEN, H. O. and RICHTER, P. H.: Homoclinic bifurcation and the asymptotic fate of periodic points. to appear

[33] PEITGEN, H. O., SAUPE, D. and SCHMITT, K.: Nonlinear elliptic boundary value problems versus their finite difference approximations: numerically irrelevant solutions, J. reine angew. Math. 322, (1981), 74-117.

[34] PEITGEN, H. O., SCHMITT, K.: Perturbations topologiques globales des problèmes non linéaires aux valeurs propres. C. R. Acad. Sc. Paris 291, (1980), 271-274.

[35] PEITGEN, H. O. and SCHMITT, K.: Positive and spurious solutions of nonlinear eigenvalue problems, in [6], 275-324.

[36] PEITGEN, H. O. and SCHMITT, K.: Global topological perturbations of nonlinear elliptic eigenvalue problems, Math. Meth. in the Appl. Sci. 5, (1983), 376-388

[37] RABINOWITZ, P. H.: Some global results for nonlinear eigenvalue problems, J. Funct. Anal. 7, (1971), 487-513.

[38] SMALE, S.: Differential dynamical systems, Bull. Amer. Math. Soc. 73, (1967), 747-817.

[39] SMOLLER, J. and WASSERMANN, A.: Global bifurcation of steady-state solutions, J. Differential Equations 39, (1981) , 269-290.

[40] STUART, C. A.: Concave solutions of singular non-linear differential equations, Math. Z. 136, (1974), 117-135.

[41] USHIKI, S.: Chaotic behavior in analytic dynamical systems, Proc. Int. Symp. on Appl. Math. and Inf. Sciences, Kyoto University, (1980).

[42] USHIKI, S.: Unstable manifolds of analytic dynamical systems, J. Math. Kyoto Univ. 21, (1981), 763-785.

[43] VOGELAERE, DE, R.: On the structure of symmetric periodic solutions of conservative systems, with applications, Contributions to the Theory of Nonlinear Oscillations, vol. 4, S. Lefschetz, ed., Princeton Univ. Press, (1968).

[44] WALTHER, H.-O.: A theorem on the amplitudes of periodic solutions of differential delay equations with applications to bifurcation, J. Differential Equations 29, (1978), 396-404.

[45] YAMAGUTI, M. and USHIKI, S.: Chaos in numerical analysis of ordinary differential equations, Physica 3D, (1981), 618-626.

[46] ZEHNDER, E.: Homoclinic points near elliptic fixed points, Comm. Pure Appl. Math. 26, (1973), 131-182.

ROGER D. NUSSBAUM

Department of Mathematics
Rutgers University
New Brunswick, New Jersey 08903

HEINZ-OTTO PEITGEN

Forschungsschwerpunkt "Dynamische Systeme"
Universitaet Bremen
2800 Bremen 33

General instructions to authors for
PREPARING REPRODUCTION COPY FOR MEMOIRS

For more detailed instructions send for AMS booklet, "A Guide for Authors of Memoirs." Write to Editorial Offices, American Mathematical Society, P. O. Box 6248, Providence, R. I. 02940.

MEMOIRS are printed by photo-offset from camera copy fully prepared by the author. This means that, except for a reduction in size of 20 to 30%, the finished book will look exactly like the copy submitted. Thus the author will want to use a good quality typewriter with a new, medium-inked black ribbon, and submit clean copy on the appropriate model paper.

Model Paper, provided at no cost by the AMS, is paper marked with blue lines that confine the copy to the appropriate size. Author should specify, when ordering, whether typewriter to be used has PICA-size (10 characters to the inch) or ELITE-size type (12 characters to the inch).

Line Spacing – For best appearance, and economy, a typewriter equipped with a half-space ratchet – 12 notches to the inch – should be used. (This may be purchased and attached at small cost.) Three notches make the desired spacing, which is equivalent to 1-1/2 ordinary single spaces. Where copy has a great many subscripts and superscripts, however, double spacing should be used.

Special Characters may be filled in carefully freehand, using dense black ink, or INSTANT ("rub-on") LETTERING may be used. AMS has a sheet of several hundred most-used symbols and letters which may be purchased for $5.

Diagrams may be drawn in black ink either directly on the model sheet, or on a separate sheet and pasted with rubber cement into spaces left for them in the text. Ballpoint pen is *not* acceptable.

Page Headings (Running Heads) should be centered, in CAPITAL LETTERS (preferably), at the top of the page – just above the blue line and touching it.

LEFT-hand, EVEN-numbered pages should be headed with the AUTHOR'S NAME;
RIGHT-hand, ODD-numbered pages should be headed with the TITLE of the paper (in shortened form if necessary).
Exceptions: PAGE 1 and any other page that carries a display title require NO RUNNING HEADS.

Page Numbers should be at the top of the page, on the same line with the running heads.

LEFT-hand, EVEN numbers – flush with left margin;
RIGHT-hand, ODD numbers – flush with right margin.
Exceptions: PAGE 1 and any other page that carries a display title should have page number, centered below the text, on blue line provided.

FRONT MATTER PAGES should be numbered with Roman numerals (lower case), positioned below text in same manner as described above.

MEMOIRS FORMAT

It is suggested that the material be arranged in pages as indicated below.
Note: Starred items (*) are requirements of publication.

Front Matter (first pages in book, preceding main body of text).

Page i – *Title, *Author's name.

Page iii – Table of contents.

Page iv – *Abstract (at least 1 sentence and at most 300 words).

*1980 Mathematics Subject Classifications represent the primary and secondary subjects of the paper. For the classification scheme, see Annual Subject Indexes of MATHEMATICAL REVIEWS beginning in December 1978.

Key words and phrases, if desired. (A list which covers the content of the paper adequately enough to be useful for an information retrieval system.)

Page v, etc. – Preface, introduction, or any other matter not belonging in body of text.

Page 1 – Chapter Title (dropped 1 inch from top line, and centered).
Beginning of Text.
Footnotes: *Received by the editor date.
Support information – grants, credits, etc.

Last Page (at bottom) – Author's affiliation.

ABCDEFGHIJ–AMS–8987654